Organic Farming: Principles and Practices

Organic Farming:
Principles and Practices

Cruz Hawkins

RCALLISTO REFERENCE

www.callistoreference.com

Callisto Reference,
118-35 Queens Blvd., Suite 400,
Forest Hills, NY 11375, USA

Visit us on the World Wide Web at:
www.callistoreference.com

ISBN: 978-1-64116-516-7 (Hardback)

Cataloging-in-Publication Data

Organic farming : principles and practices / Cruz Hawkins.
    p. cm.
Includes bibliographical references and index.
ISBN 978-1-64116-516-7
1. Organic farming. 2. Agriculture. 3. Agricultural systems. I. Hawkins, Cruz.
S605.5 .O74 2022
631.584--dc23

Table of Contents

Permissions

Index

Preface

The farming practice which primarily focuses on the use of natural organic substances and prohibits the use of synthetic substances is known as organic farming. It advocates the usage of organic products due to reasons related to sustainability, self-sufficiency, food security and safety. Organic farming is an alternative agricultural system that uses organic fertilizers such as compost manure, green manure and bone meal. It involves the use of techniques such as crop rotation and companion planting. It also incorporates biological pest control and mixed cropping as well as fostering of insect predators for the management of pests. Methods of organic farming are studied within the field of agroecology. This textbook contains some path-breaking studies in the field of organic farming. It strives to provide a fair idea about this discipline and to help develop a better understanding of its varied principles and practices. Those in search of information to further their knowledge will be greatly assisted by this book.

A foreword of all chapters of the book is provided below:

Chapter 1 - Organic farming is an alternative agriculture system that uses organic fertilizers and techniques such as crop rotation and companion planting. Organic farming methods are the combination of scientific knowledge of ecology, traditional farming practices and modern technology. This is an introductory chapter which will introduce briefly all the significant aspects of organic farming such as its importance as well as the different methods used in it.; **Chapter 2** - There are a number of types of organic farming such as organic aquaculture, organic horticulture and organic gardening. Vegan-organic farming is a type of organic farming which seeks to cause minimal harm to animals. This chapter discusses in detail the techniques and methods used in these types of organic farming.; **Chapter 3** - Pest management in organic farming is done by manipulating agroecosystem processes. Weed control is an important part of pest control that aims to stop weeds from growing in desired flora and fauna. This chapter discusses in detail the theories and methodologies related to organic weed and pest management.; **Chapter 4** - The fertilizers that are derived from organic sources such as human excreta, animal matter, animal excreta and vegetable matter are known as organic fertilizers. Peat, manure and animal waste are some of the naturally occurring organic fertilizers. This chapter has been carefully written to provide an easy understanding of the varied facets of organic fertilizers as well as the processes used in their production, such as vermicomposting.; **Chapter 5** - Diseases in plants can be caused due to pathogens such as fungi, bacteria and viruses. There are various organic methods for disease management such as planting resistant cultivars, exclusion and crop rotation. All these diverse organic methods for disease management have been carefully analyzed in this chapter.

At the end, I would like to thank all the people associated with this book devoting their precious time and providing their valuable contributions to this book. I would also like to express my gratitude to my fellow colleagues who encouraged me throughout the process.

Cruz Hawkins

Chapter 1

Understanding Organic Farming

Organic farming is an alternative agriculture system that uses organic fertilizers and techniques such as crop rotation and companion planting. Organic farming methods are the combination of scientific knowledge of ecology, traditional farming practices and modern technology. This is an introductory chapter which will introduce briefly all the significant aspects of organic farming such as its importance as well as the different methods used in it.

Organic farming is a form of agriculture production techniques where plants and animals are grown in a natural way which means growing food in harmony with nature. This process involves avoiding the use of synthetic materials but using biological materials which are available in the natural habitat in growing crops and animal husbandry. In other words it's a science of using natural techniques in growing crops and livestock without harming the natural environment.

It is integrating the modern farming technology with traditional farming methods of past which are still useful today. Organic farming system rely upon using of animal waste, green manures, crop residue, rotation of crops, organic wastes, bio fertilizers, vermicompost, and bio pest controllers to maintain the soil health and productivity, supplementing plant nutrients, and controlling insects, pests, and weeds. In contrast to modern agriculture, organic farming avoids using of synthetic fertilizers, pesticides, herbicides, growth promoters or regulators, and livestock feed additives.

The methods used in organic farming makes the soil self-sufficient while increasing farm productivity and repairs the abused farm fields and lands by massive usage of chemical fertilizers and toxic pesticides in agriculture sector. Many of the farmers who did not follow nature's holistic farming ways have decreased soil fertility, loss of topsoil, water contamination on both ground and surface and loss of genetic diversity.

Basic Principle of Organic Farming and Agriculture

The basic principle of organic farming is to support the general idea that the soil, plants, animals and man are connected. The role of organic farming in agriculture practices to sustain and enhance the health of both land and environment from smallest organisms in the soil to the top most in the food chain. It must conduct and balance natural and environmental resources for production and consumption in all fairness to human beings and to all living beings in their natural behavior and well-being. Precaution, care and responsibility are the key concerns to ensure for healthy, safe and ecologically sound organic practices are passed for the future generations.

The important role of organic farming is to contribute successful management of agricultural resources and enhancement of sustainability. There must be a balance to satisfy human needs through agricultural resources while at the same time preserving natural resources for the future. Many organic farms have produced higher yields than conventional farming methods. Soil health in the organic farms has significant rates of nitrogen, minerals, nutrients, and microbial abundance and diversity are found. The following are the organic farmer's methods in utilizing materials.

- Preserve and build good soil structure and fertility by recycling and composting crop wastes and animal manures.

- Encouraging polyculture against monoculture cultivation. Farming a single crop may reduce cost on.

- fertilizers, seeds and pesticides but on a long run creates problem such as reduction in soil fertility and soil erosion. Whereas in monoculture, growing variety of crops attract different soil microbes and certain crops act as pest repellant.

- Mixed farming of crop and livestock production.

- Control pests, weeds and diseases through careful planning and crop choice. Use natural pesticides and allowing useful predators that eat pests.

- Help in careful use of water resources.

- Good animal husbandry.

Methods of Organic Farming

Growing crops by using eco-friendly methods and avoiding man made farming products such as chemical fertilizers, toxic pesticides and insecticides, growth regulators and genetic modification of crop species is the best farming practice.

Simple eco-friendly farming methods use various traditional agriculture practices using of compost fertilizers, mixed farming, crop rotation, biological pest control, minimum tillage. The use of these methods will enhance the crop productivity without synthetic materials and a sustainable farming approach.

Following are the methods practiced in organic farming.

1. Soil fertility in organic farming systemsrely on the management of soil organic matter. To optimize crop production the soil organic matter is to be enhanced in chemical, biological and soil physical properties. Farmers test soils at most testing labs to fertilize the crop but not the soil. Farmer must learn to feed the soil in turn the soil will feed the crop. Living microbes and other organisms are the contributors to soil fertility on a sustained basis. They conserve the fertility and top soil erosion by implementing appropriate conservation practices. Soil fertility can be boosted by incorporating old decomposed farm yard manure, composts and legumes. Soils health can be measured by its aggregate stability (tilth), water retention capacity, drainage, porosity, bulk density, resistance to crusting and compaction.

2. Crop diversity through crop rotation is growing of variety of crops in same land also called polyculture. An organic farmer must understand crop variety and its specific needs such as type of soils, rainfall, climatic conditions, and growing duration. Crop rotation build better soil, control pests, and gives farm profits. A farmer has to include cash crops, filler crops, and cover crops in a

season through crop rotation. Crops must be grown in a specific order over a time and the cycle will be repeated over a period of years.Crop diversity through deliberate crop rotation prevents the buildup of pest population by interrupting pest habitats and its life cycle. Farm land or farm beds may be rotated for weed or pest control and sometimes by land tillage.

3. Nutrient management in organic farming is one of the important key factors in crop production. Farmer must understand soil is a living system and agricultural productivity is directly dependent on soil fertility. The soil surrounding the root zones in which bacterial grow must be stimulated, these microbes or microorganisms in their turn supply nutrients to the roots. Crop production and nutrient cycles must be balanced as soil health is exploited by cultivation which leads to soil degradation through nutrient depletion. On a long term planning avoid chemical fertilizers, growth boosters but by using organic manures, bio fertilizers, and compost. Farmers must learn to rely on long term soil quality for productivity of fields than relying on chemical fertilizers and pesticides. The organic sources add different nutrients to the soils while resists soil erosion, and holds water. Plants that need some natural minerals to grow can also be added. Soil pH balance can be maintained by adding lime. There are different ways to increase soil nutrients.

4. Organic manures: Farmyard manure is naturally available organic manure. It is a mixture of animal excreta, leaves, straw, and grass. Poultry manure is also one of the extremely valuable organic manure rich in phosphorus. The collected animal waste must be stored in aerobic conditions till the manure shows a brown to black colour.

5. Bio fertilizers: Is a substance that contains living microorganisms that colonizes at the root zone or rhizoshpere by increasing supply of nutrients in the soil. There are different types of bio-fertilizers available such as Rhizobium, Azotobacter, Azospirillum, Cyanobacteria, Plant growth promoting rhizobacteria, Phosphorus-solubilizing bacteria (PSB), Mycorrhizal fungi, Blue green algae (BGA), and Azolla.

6. Mineral Fertilizer: Are used as supplement to organic manures. They control the soil pH, soil salinity and must be used as recommended after soil test. Mineral fertilizers are based on ground natural rock. They can be obtained from stone powder, rock phosphate, lime, and plant ashes.

7. Composting: This is a process of recycling decomposing of organic materials such as crop residue, plants, leaves, twigs into a rich soil. By composting once any living thing, organic waste returns back into soil delivering nutrients completing a cycle of life to continue.

There are three types of composting:

- Backyard composting.

- Worm composting.

- Grass cycling.

Green manures- Using of green plant tissues to improve the soil fertility and physical structure by incorporating them into soil. The important aspect of green manuring is enriching soil with ecomposed organic matter into the soil. This can be done in two ways. Firstly by growing manure crops belonging to leguminous family and secondly by collecting twigs with green leaf from telands.

Some of the important green manure crops are sunnhemp, urd, dhaincha, cowpea, sbaniarostrata, mung, cluster beans, shervi, and pillipesara.

Managing Weeds in Organic Farming

Weed management is one of the major issues in organic farming. The important factors know about its reproduction and disposal strategies in determining their spread and to control. Weeds are unwanted plants having heavy seeding perennials with persistent underground parts such as stolons and rhizomes becoming a problematic in farm production. While some of the heavy seeding annuals may be effectively controlled by avoiding favourable weed germination and growth conditions. Much of the weed controlling is done by manual mowing, discing, plowing or hand pulling of weeds as in organic farming chemical herbicides is avoided. The other practices to control weeds include tillage, flooding, mulching or by restricting water supply through drip irrigation.

Mulching in Organic Farming

Mulching is a process to cover the surface of the soil with any decomposable materials such as straws, hay, leaves, crop residues, grass etc. Mulching controls the soil moisture, temperature which enhances the activity of earthworms and other soil organisms. Earthworms help creating small and large pores in the soil structure that lets aeration and water in the soil. Mulch suppress growth of weed and decomposed mulch will add nutrients to the soil.

Pest and Disease Management in Organic Farming

- Plants must not be exposed to too much or too less sunlight.

- Plants must not be exposed to low or strong heat and temperature.

- Plants must not be exposed to less irrigation or water logging.

- Plants must not be exposed to excess nutrients or nutrient deficiency.

Prevention practices must be followed to keep the plants healthy. A farmer must have enough knowledge about pests, diseases to choose effective preventive measures. To protect crop suitable combination of different methods at the right time must be under taken. Some of the important preventive measures are given below:

- Choose right cultivars which are disease resistant and well adapted to local climatic conditions.

- Choose seeds that are clean and safe from pathogens and weeds.

- Mixed farming or poly-culture will limit pest and disease.

- Practice moderate and balanced field nutrient management.

- Apply suitable soil cultivation methods.

- Proper irrigation management ensures no water logging.

- Promote natural enemies by providing natural habitat.

- Planting at optimum time and spacing.

- Proper disposal measures of infected plant parts and residue from field without spreading.

Livestock in Organic Farming

Organic farming encourages keeping farm animals as part of the farming system to increase sustainability of the farm. Farming system depends on the internal flow of nutrients to the soil from animal waste there by producing crops and the crop residue is available as feed to the livestock in the farm thereby forming a cycle. Therefore, the waste product of one becomes a resource for the other. Farmers must have sufficient information and knowledge about the integration of crops and livestock which will be economically and environmentally sustainable allowing the maximum use of available resources over a long period.

Genetic Modification in Organic Farming

Genetically modified crops or plants have high resistance to disease and high yields by altering their DNA. There is a risk on consuming genetically modified foods both human health and to our environment. Organic farming keeps away this kind of agriculture techniques and focuses on using natural ways by discouraging genetically modified livestock and crops.

Advantages of Organic Farming

- Organic manures and compost in still soil health thereby increasing quality and crop yield.

- The supply of nutrients and micronutrients to the soil are optimum.

- It reduces agriculture procuring cost needs.

- Organic farming helps in preventing the environment damage and degradation of the soil.

- Organically grown plants are more resistant to pests and diseases.

- Organic manures are mostly the by-products of animals and plants.

- Biological diversity in organic farming produces crop yields, milk, eggs, and meat.

- Slows down global warming and fewer residues in food.

- Fruits and vegetables produced organically have good flavor with rich nutrition values.

Disadvantages in Organic Farming

- More manual work or labor is required to carry the hard work for successful production.

- Organic products are more expensive in the market.

- Marketing of organic products under certification is very expensive.

- The biggest disadvantage in organic farming method is time consuming.

- Farmer must have proper knowledge and skill in choosing alternative of chemicals.

Successful Organic Transition

The transition from conventional to organic farming requires numerous changes. One of the biggest changes is in the mind-set of the farmer. Conventional approaches often involve the use of quick-fix remedies that, unfortunately, rarely address the cause of the problem. Transitioning farmers generally spend too much time worrying about replacing synthetic input with allowable organic product instead of considering management practices based on preventative strategies. Here are a few steps new entrants should follow when making the transition to organic farming:

Basics of Organic Agriculture and the Organic Farming Standards

Since organic production systems are knowledge based, new entrants and transitional producers must become familiar with sound and sustainable agricultural practices. Transitional producers should be prepared to read appropriate information, conduct their own trials and participate in formal and informal training events. As mentioned, switching from conventional to organic farming is more than substituting synthetic materials to organic allowed materials. Organic farming is a holistic system that relies on sound practices focused on preventative strategies. Since there are often few organic remedies available to organic producers for certain problems, prevention is the key element in organic production.

Plan your Transition Carefully

Develop a transitional plan with clear and realistic goals. The plan should clearly identify various steps to be taken in making the transition to organic and be sure to include realistic timeframes. Identify your strengths and weaknesses. Consider ways to address any weaknesses, while building on strengths. The business side of the transitional plan should contain a multiple year budget and an effective/realistic marketing strategy. Make sure your list of expenses is comprehensive. Include all prerequisites to begin the transition; such as, mechanical weeding equipment, specialized composting equipment and applicators, additional handling equipment dedicated to the organic products, and processing equipment. Although the demand for organic products is continually growing, growers need to make sure they have a reliable market for the organic products they plan to produce.

Careful planning is very important. During the early part of the transitional period, yields are often depressed and premium prices for certified organic products are generally not yet obtainable. Use realistic yields and prices when evaluating the feasibility of your project.

In some instances, it is preferable to continue using conventional measures early on in the transitional process in order to avoid dramatic yield reduction which could jeopardize the financial well-being of the operation. Farmers who are planning to convert their livestock operation should consider certifying their fields first. This allows time to learn more about organic livestock management requirements while, at the same time, starting to produce organic feeds.

Parallel production is the simultaneous production, processing or handling of organic and non-organic crops, livestock and other products of a similar nature. Although this type of activity is highly discouraged by certifiers, some allow it, especially during the transition period. If permitted to practice parallel production, producers must be prepared to deal with significant record keeping in order to ensure traceability and organic integrity.

Understand your Soils and Ways to Improve them

Since soil is the heart of the organic farming system, it is crucial that new entrants understand the various characteristics and limitations of the soils found on their farm. Soil suitability may vary significantly from one field to the next. Fields with good drainage, good level of fertility and organic matter, adequate pH, biological health, high legume content, and with less weed and pest pressure, are excellent assets. Often these fields are the first ones ready for transition and certification.

Many tools exist to assess soils. Soil chemical, physical and biological analyses, soil survey and legume composition field assessments, and field yield histories are very important and should be considered early in the transition. Unhealthy soils require particular attention.

If farmers plan to grow crops without raising any livestock, it may be necessary for them to source allowable soil amendments such as composted manure, limestone, rock dust, and supplementary sources of nitrogen, phosphorus, potassium and micro-nutrients. Even with the best of crop rotations that include green manure crops like legumes (nitrogen fixing crops), transitional growers will be challenged if they want to obtain optimal yields without additional livestock manure, compost and/or other off-farm soil inputs. When these inputs are scarce or expensive, producers may benefit from integrating livestock on their farm.

Let's not forget, under organic production, farmers must be able to recycle nutrients through proper nutrient management practices: recycling through good manure and compost utilization, crop rotations, cover crops (green manure, catch, and nitrogen fixing crops), and by reducing nutrient losses due to leaching, over-fertilization, as well as poor manure and compost management (storage, handling, and spreading).

Identify the Crops or Livestock Suited for your Situation

Before growing a crop or raising any livestock, consider the following: degree of difficulty to grow or raise the product organically, land and soil suitability, climate suitability, level of demand for the product, marketing challenges, capital required, current prices for conventional, transitional and organic products, and profitability over additional workload.

Design Good Crop Rotations

Once the crops are chosen, carefully plan the crop rotation(s) and select the most suitable cover crops (green manure, winter cover crops, catch crops, smother crops, etc.). Crop rotations are extremely important management tools in organic farming. They can interrupt pest life cycles, suppress weeds, provide and recycle fertility, and improve soil structure and tilth. Some rotational crops may also be cash crops, generating supplemental income.

On some farms, land base availability may be a limiting factor when planning your crop rotations. The transitional plan should, therefore, include crop rotation strategies. Responding to external forces such as new market opportunities may also have a significant impact on crop rotations, so farmers need to consider the effect that growing new crops has on their crop rotations and land base availability.

Identify Pest Challenges and Methods of Control

It is important to know the crop's most common pests, their life cycles and adequate control measures. For instance, Colorado potato beetle may be a pest of significant importance when growing potatoes; cucumber beetles in cucurbitaceous crops (cucumber, squash, and melons); flea beetle in many seedlings crops; clipper weevil and Tarnish Plant Bug in strawberry crops.

There are several measures available to reduce pest pressure: crop rotation, variety selection, sanitation, floating row covers, catch crops, flamers, introduction of beneficial insects, bio pesticides, and inorganic pesticides. Transitional growers should be prepared to use and experiment with some of these options. When considering a new type of production, discuss pest issues with your agrologists, IPM specialists and other existing organic producers to optimize your chances of success.

Availability of organic supplies has improved significantly over the past few years. New pest control products containing B.t., spinosad, kaolin clay are effective and currently available to organic growers. It is often reported that the types of weeds found on the farm evolve with time as growers change the way they grow their crops and control their weeds. By keeping track of the weed population, growers will be able to refine their crop rotations and improve their control measures.

Under organic livestock management, cattlemen must provide attentive care that promotes health and meets the behavioral needs of various types of livestock. With good herd health practices, farmers rarely need to rely on conventional medicine. Organic cattlemen should, however, try to familiarize themselves with alternative remedies such as herbal/aroma therapies, homeopathy, and immune system promoters.

Be Ready to Conduct your Own On-farm Trials

Successful organic farmers continuously try new and innovative management practices. Practices such as cover cropping, inter-planting, and use of various soil and pest control materials need to be evaluated regularly by organic farmers. Be prepared to try new approaches.

Be Ready to keep Good Records

Record keeping is one of the most important requirements to maintain organic integrity. Farmers are expected to keep detailed production, processing and marketing information. This information includes everything that enters and exits the farm. Third party, independent inspectors require farmers to present the above mentioned documentation when inspecting the farm operation. Once the record-keeping requirements are understood and the reporting procedure established, paperwork becomes routine.

Avoid these Common Mistakes

- Underestimating the need for good transitional and marketing plans.

- Underestimating the need to fully understand the Organic Standard. Organic producers must understand the standard in order to know what is permitted and prohibited.

- Failing to think prevention. Transitional farmers should consider improving their crop rotation, soil and crop management skills, livestock management practices (feeding program, heard health program, grazing system, housing facilities, and husbandry).

Particulars	Conventional farming	Organic farming
Application of compost/FYM	√	√
Judicious application of inorganic fertilizers	√	×
Biofertilizers	√	√
Pesticide applications	√	×
Fungicide applications	√	×

Principles of Organic Agriculture

The Principle of Health. The Principle of Ecology. The Principle of Fairness. The Principle of Care.

The four principles of organic agriculture: Health, Ecology, Fairness & Care.

These principles are the roots from which Organic Agriculture grows and develops. They express the contribution that Organic Agriculture can make to the world. Composed as inter-connected ethical principles to inspire the organic movement in its full diversity, they guide our development of positions, programs and standards.

Principle of Health

Organic Agriculture should sustain and enhance the health of soil, plant, animal, human and planet as one and indivisible.

This principle points out that the health of individuals and communities cannot be separated from the health of ecosystems - healthy soils produce healthy crops that foster the health of animals and people.

Farmer sifting compost Man feeding two steer

Health is the wholeness and integrity of living systems. It is not simply the absence of illness, but the maintenance of physical, mental, social and ecological well-being. Immunity, resilience and regeneration are key characteristics of health.

The role of Organic Agriculture, whether in farming, processing, distribution, or consumption, is to sustain and enhance the health of ecosystems and organisms from the smallest in the soil to human beings. In particular, organic agriculture is intended to produce high quality, nutritious food that contributes to preventive health care and well-being. In view of this it should avoid the use of fertilizers, pesticides, animal drugs and food additives that may have adverse health effects.

The Principle of Ecology

Organic Agriculture should be based on living ecological systems and cycles, work with them, emulate them and help sustain them.

Seedlings thriving in the sun

This principle roots Organic Agriculture within living ecological systems. It states that production is to be based on ecological processes, and recycling. Nourishment and well-being are achieved through the ecology of the specific production environment. For example, in the case of crops this is the living soil; for animals it is the farm ecosystem; for fish and marine organisms, the aquatic environment.

Organic farming, pastoral and wild harvest systems should fit the cycles and ecological balances in nature. These cycles are universal but their operation is site-specific. Organic management must be adapted to local conditions, ecology, culture and scale. Inputs should be reduced by reuse, recycling and efficient management of materials and energy in order to maintain and improve environmental quality and conserve resources.

Organic Agriculture should attain ecological balance through the design of farming systems, establishment of habitats and maintenance of genetic and agricultural diversity. Those who produce, process, trade, or consume organic products should protect and benefit the common environment including landscapes, climate, habitats, biodiversity, air and water.

The Principle of Fairness

Organic Agriculture should build on relationships that ensure fairness with regard to the common environment and life opportunities.

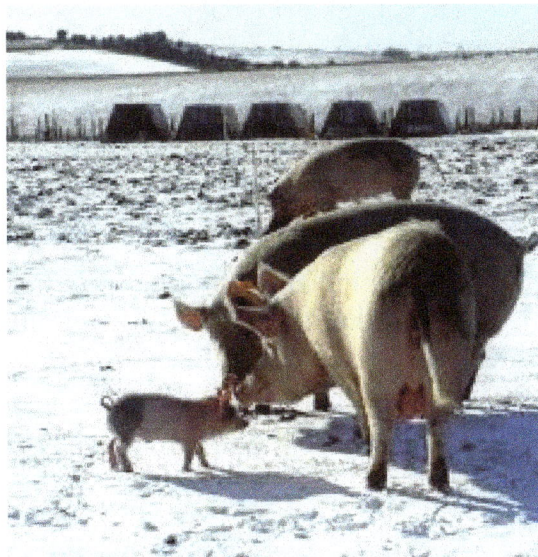

Pigs in the snow

Fairness is characterized by equity, respect, justice and stewardship of the shared world, both among people and in their relations to other living beings.

This principle emphasizes that those involved in Organic Agriculture should conduct human relationships in a manner that ensures fairness at all levels and to all parties - farmers, workers, processors, distributors, traders and consumers. Organic Agriculture should provide everyone involved with a good quality of life, and contribute to food sovereignty and reduction of poverty. It aims to produce a sufficient supply of good quality food and other products.

This principle insists that animals should be provided with the conditions and opportunities of life that accord with their physiology, natural behavior and well-being.

Natural and environmental resources that are used for production and consumption should be managed in a way that is socially and ecologically just and should be held in trust for future generations. Fairness requires systems of production, distribution and trade that are open and equitable and account for real environmental and social costs.

The Principle of Care

Organic Agriculture should be managed in a precautionary and responsible manner to protect the health and well-being of current and future generations and the environment.

Organic Agriculture is a living and dynamic system that responds to internal and external demands and conditions.

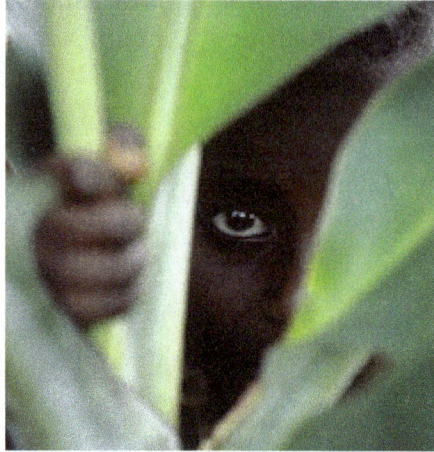
Child peeking through tall plants

Practitioners of Organic Agriculture can enhance efficiency and increase productivity, but this should not be at the risk of jeopardizing health and well-being. Consequently, new technologies need to be assessed and existing methods reviewed. Given the incomplete understanding of ecosystems and agriculture, care must be taken.

This principle states that precaution and responsibility are the key concerns in management, development and technology choices in Organic Agriculture.

Science is necessary to ensure that Organic Agriculture is healthy, safe and ecologically sound. However, scientific knowledge alone is not sufficient. Practical experience, accumulated wisdom and traditional and indigenous knowledge offer valid solutions, tested by time.

Organic Agriculture should prevent significant risks by adopting appropriate technologies and rejecting unpredictable ones, such as genetic engineering. Decisions should reflect the values and needs of all who might be affected, through transparent and participatory processes.

Basic Methods of Organic Farming

Pest control

Organic farming is done to release nutrients to the crops for increased sustainable production in an eco-friendly and pollution-free environment. It aims to produce crop with a high nutritional value. There are various methods by which organic farming is practiced are as follows:

1. Crop Diversity: Now-a-days a new practice has come into picture which is called -Polyculture- in which a variety of crops can be cultivated simultaneously just to meet the increasing demand of crops. Unlike the ancient practice which was -Monoculture- in which only one type of crop was cultivated in a particular location.

2. Soil Management: After the cultivation of crops, the soil loses its nutrients and its quality depletes. Organic agriculture initiates the use of natural ways to increase the health of soil. It focuses on the use of bacteria that is present in animal waste which helps in making the soil nutrients more productive to enhance the soil.

3. Weed Management: Weed, is the unwanted plant that grows in agricultural fields. Organic agriculture pressurizes on lowering the weed rather than removing it completely.

4. Controlling other organisms: There are both useful and harmful organisms in the agricultural farm which affect the field. The growth of such organisms needs to be controlled to protect the soil and the crops. This can be done by the use of herbicides and pesticides that contain less chemicals or are natural. Also, proper sanitization of the entire farm should be maintained to control other organisms.

5. Livestock: Organic farming instigates domestic animals use to increase the sustainability of the farm.

6. Genetic Modification: Genetic modification is kept away from this kind of agricultural set up because organic farming focuses on the use of natural ways and discourages engineered animals and plants.

Cover Crop Species for Organic Farming Systems

The diversity of cover crop species and numerous planting combinations create opportunities for nearly every farming system to include cover crops.

Main Groups of Cover Crops

Cover crops are generally grouped according to their botanical classification. Two groups of plants, grasses and legumes, account for most of the plant species used as cover crops. However, there are many other excellent crops that offer a range of benefits to a farming system that do not fall into either of these groups including: buckwheat (Fagopyrum esculentum), a member of the buckwheat family (Polygonaceae); sunflower (Helianthus spp.), a member of the aster family (Asteraceae); and a wide variety of mustards (Brassica spp.) and forage radishes (Raphanus sativus) that are members of the crucifer family (Brassicaceae).

The grass family (Poaceae) of cover crops include grains and forage grasses such as cereal rye (Secale cereale L.), sorghum [Sorghum bicolor spp. bicolor], sorghum-sudangrass hybrids (Sorghum bicolor X S. bicolor var. sudanense), and wheat (Triticum aestivum). These crops are often grown to produce income, but they also make excellent cover crops due to the many ecological benefits they provide to the farming system. They establish quickly (an important characteristic for good weed suppression), produce high biomass and dense fibrous root systems (to prevent soil erosion, sequester carbon, increase soil organic matter content, and improve soil quality), and can scavenge and store available soil nutrients such as nitrogen (N) and potassium (to reduce nutrient losses due to leaching). Some grasses, especially rye, sorghum, and sorghum-sudangrass, further suppress weeds through the release of natural substances that inhibit the growth of neighboring plants, a process called allelopathy. Allelopathic substances from rye, sorghum-sudangrass, and other cover crops have been shown to inhibit the emergence and early growth of many weeds.

Legumes are a large group of plants in the bean family (Fabaceae) that includes vetch (Vicia spp.), clover (Trifolium spp.), field peas (Pisum sativum), cowpea (Vigna unguiculata), and soybean (Glycine max). Most plants in this group have a beneficial relationship with specific soil bacteria (Rhizobium spp.). The bacteria form nodules and live in the plant's roots, and convert N_2 gas from the surrounding air to plant-available nitrate in a process called nitrogen fixation. Thus, legumes are grown primarily to provide a source of N to a subsequent crop and are most beneficial when the soil N levels are low. Legumes can produce substantial biomass, attract beneficial insects, and suppress weeds through competition and, in some cases, allelopathy.

Single Species vs. Mixtures

Single species of cover crops are often planted if operational constraints limit selection to a single species or when a farmer's objective is known to be best fulfilled by a specific plant species. For example, some cover crops such as sunn hemp, a summer-planted legume, suppress particular species of soil pathogenic nematodes. The cover crop creates an environment below the soil that is unfavorable for nematodes. Nematodes produce less young when they are under stress. Over time, the unfavorable environment results in a reduction of active nematodes in the soil. Other advantages of planting a single species of cover crop include ease of planting and uniform establishment and maturity. Disadvantages include increased risk of poor establishment and development due to adverse weather conditions or pests, and a more limited range of benefits to the cropping system. In addition, a grain or grass cover crop monoculture may tie up soil N after the cover crop is tilled in or otherwise terminated, while a legume monoculture may release soluble N so rapidly that much of it leaches before the subsequent crop can utilize it.

Mixtures of cover crop species are planted to optimize carbon to nitrogen (C:N) balance, obtain multiple benefits, or more fully achieve a particular objective such as organic matter production or weed suppression. At maturity (flowering), a grass-legume cover crop biculture often contains organic carbon and nitrogen in a ratio of 25:1 to 30:1, which promotes a slow release of N that the subsequent crop can utilize efficiently. Fresh organic residues with C:N ratios in this range also form the most stable humus as they decompose. In addition, planting a mixture of cover crops can reduce risk of crop failure; although it can require additional planning and labor.

Figure: Spring oat and Austrian winter field pea mixture, planted in early April, photographed in mid-June. The cover crop has reached 3 ft tall and 3 tons/acre aboveground biomass

Allelopathy is known to be a species-specific phenomenon. Thus, planting mixtures of cover crops can enhance the cover crop's allelopathic potential by achieving a broader spectrum of weed control because each cover crop species contributes allelopathic activity towards a specific weed species.

Mixtures can also be planted to influence insect populations. Cover crops can be selected to attract beneficials or deter pests. Like the previous examples, many of these cover crop-insect relationships are species specific, so a little homework can go a long way in making sure the objectives are achieved.

Two commonly planted grass and legume mixtures are:

- Cowpea and sorghum: Above-ground biomass and the nutrients in that biomass can be increased because this mixture can utilize more below-ground and above-ground niches for nutrients, water, and light than the monocultures alone. Sorghum is a very heavy N feeder with a tall, upright growth habit, while cowpea fixes N and forms a dense canopy between the sorghum stalks. Together, the allelopathic sorghum and densely-shading cowpea deliver a one-two punch on weeds.

- Hairy vetch (Vicia villosa) and cereal rye: Rye and other cereal crops usually germinate and establish effective root systems more rapidly than legumes and lower the soil N concentration. Since nodulation of legume roots and fixation of atmospheric N by legumes is generally greater when the soil N concentration is low, nodulation and N fixation is increased in mixtures. The web of fibrous rye roots helps prevent frost-heaving damage to overwintering vetch, and the tall, strong rye stems provide support for vetch vines during rapid spring growth. Biomass production and weed suppression by the combination is often better than from either crop alone.

Selecting the best cover crop species or combination of cover crop species for a particular climate, soil, and farming system may take a bit of trial and error.

Importance of Organic Farming

Modern farming practices heavily based on use of chemical fertilizers and pesticides, have created problems of land degradation, environmental pollution, deforestation, biodiversity depletion, seepage and water logging, lowering of ground water tables, inter crop disparities, emergence of several diseases, pest multiplication, pest resurgence and resistance.

Modern agricultural practices are the major cause of soil, water and air pollution. Chemical fertilizers crowd out useful minerals naturally present in the topsoil. The microbes like bacteria, fungi, actinomycetes, worms etc. in top soil enrich the humus and help to produce nutrients to be taken up by the plants and later by animals. However, fertilizer enriched soil is unable to support microbial life and hence there is less humus and less nutrients and the soil easily becomes poor and eroded by rain and wind.

Excessive uses of nitrogenous fertilizer in modern farming decreases the potassium content of crops. Similarly excessive potash treatment decreases valuable nutrients in foods, such as ascorbic acid and carotene. The use of superphosphate leads to copper and zinc deficiency in crop plants. Nitrate fertilizer increase the crop yield (carbohydrate) but at the expense of proteins. Excessive fertilizer use produces over-sized fruits and vegetables, which are prone to insects and other pests.

The fertilizers used to raise the crop yield are drained by rain water to the adjacent fresh water bodies like rivers, lakes and ponds causing nutrient enrichment (especially nitrate and phosphate) of the aquatic bodies. This phenomenon is called as 'eutrophication', which triggers the luxuriant growth of blue green algae (cyanobacteria). The algal growth forms floating scums or blankets of algae called as algal blooms. Blooms of algae are generally not utilized by zooplanktons. The algal blooms compete for light for photosynthesis with other aquatic plants. Thus oxygen is depleted.

These blooms also release some toxic chemicals, which deteriorate the water quality. The decomposition of blooms also leads to oxygen depletion in water. Thus in poorly oxygenated water with higher carbon dioxide (CO_2) levels, fishes and other animals begin to die and clean water is turned into a stinking drain. The drinking of nitrate and nitrite contaminated water causes the disease 'methaemoglobinaemea' in children, which interferes with the oxygen carrying capacity of the blood. This leads to various disorders like damage to respiratory and vascular system, blue colouration of skin and even cancer.

The pesticides moving from crop fields to aquatic bodies affect the aquatic flora and fauna. Many of non-biodegradable pesticides (Chlorinated hydrocarbons) like D.D.T (Dichloro diphenyl trichloroethane), B.H.C (Benzene hexachloride) etc. enter the food chain and reach the human body causing harmful effect to human health.

The concentration of the pesticides increases with increasing food chain and the phenomenon is known as biological magnification.

The excessive use of nitrogenous fertilizers causes acidification of soil, resulting in the loss of soil fertility. The indiscriminate use of pesticides to control pests kills several useful flora and fauna of the soil, which promote soil fertility. Besides the targeted insects, useful insects promoting cross and self-pollination are also killed. This leads to decline in crop productivity owing to reduced rate of pollination accomplished by insects.

Microbial action on nitrogenous fertilizer in the soil leads to formation of nitrous oxide, which causes the thinning of stratospheric ozone layer. The latter is a protective shield filtering harmful ultraviolet radiation emanating from the sun. Excessive use of water for irrigation in modern agriculture leads to water logging which causes anaerobic condition in the soil resulting in the production of methane gas by methanogenic bacteria. Methane (CH_4) is a potent greenhouse gas causing global warming is increasing at the rate of 1% per annum.

In modern farming practices there is constant use of some high yielding varieties of the crops in place of nutritive indigenous varieties resulting into uniformity in biodiversity. This poses threat to the loss of biodiversity. The gradual loss of variability in the cultivated forms and in their wild relatives is referred to as 'genetic erosion'. This variability arose in nature over an extremely long period of time, and if lost would not be reproduced during a short time period. In modern farming, the loss of biological diversity is enhanced due to overexploitation of natural resources, excessive use of pesticides and environmental pollution.

The various aforesaid side effects of expensive modern farming on soil, crops and human health have compelled to look for an alternative in the form of organic farming. The latter is inexpensive, sound, safe and sustainable in long run without any adverse impact on the environment. Therefore, the organic farming has become inevitable to tackle with the problem of land degradation, environmental pollution, biodiversity depletion and contamination of food grains from pesticide residues.

Management Practices in Organic Farming

The management practices for organic farming differ from those of modern farming. The important steps in this type of farming are conservation of soil and genetic resources, integrated nutrient management, integrated weed management and integrated pest management.

Tillage practices in organic farming aim at reducing soil degradation. Therefore, conservational tillage is adopted in place of conventional tillage. Conservational tillage is disturbing the soil to the minimum extent necessary and leaving crop residues on the soil. Zero tillage and minimum tillage are the types of conservational tillage which reduce soil loss up to 99% over conventional tillage. In most cases, conservation tillage reduces soil loss by 50% over conventional tillage. Moreover, conservational tillage maintains the organic matter content of the soil and prevents the removal of nutrients from soil through rainwater. Conservational tillage also causes an increase in microbial and earthworms population in the soil.

Organic farming emphasizes on the cultivation of different indigenous nutritive local varieties of crops in place of few high yielding hybrid varieties only.

In organic farming, besides manure, green manure, compost and vermi-compost, oil cakes and oil meals play a key role as natural fertilizers. The commonly used organic nitrogenous fertilizers include rapeseed, mustard, neem, castor, mahua, karanja and linseed cakes. In addition to these, cakes from sal, groundnut and soyabean are also used in various combinations to increase yield and control pests.

Organic fertilizers have a slower action but they supply available nitrogen over a longer period of time. Besides, they protect useful flora and fauna of the soil; ameliorate yields and quality of products. Since there is increase in soil fertility, the biological activity is maintained intact.

In organic farming nitrogenous bio-fertilizers like Azolla pinnata (Pteridophyte), Anabaena, Aulosira, Nostoc, Scytonema, Tolypothrix, Cylinderospermum, Camptylonema, Westiellopsis (Blue green algae) and Azorhizobium, Bradyrhizobium, Mesorhizobium, Rhizobium, Sinorhizobium, Azotobacter Azospirillum and (Bacteria) are used to raise the fertility of soil. Furthermore, the fungi Aspergillus, Penicillium and Trichoderma are used as cellulolytic bio-fertilizers to enhance the rate of organic matter decomposition for the quick release of nutrients in the soil. The bacteria Bacillus subtilis, Pseudomonas putida and Pseudomonas fluorescens are used as phosphatic bio-fertilizers to solubilize phosphate.

The application of bio-fertilizers like Azospirillum makes the plant hardier by producing certain phenolic substances and eventually provides resistance to pest and diseases. Blue green algae not only fix atmospheric nitrogen but also excrete vitamin B12, ascorbic acid and auxins, which improve the growth of crop plants. They also possess the properties of solubilizing the bound phosphate of the soil.

In organic farming the weeds are controlled by employment of physical, cultural and biological method. Insect pests are controlled by combination of cultural and biological methods and at the same time use of resistant crop varieties.

Crop rotation practices are key to success of organic farming. Crop rotation is not only important for the soil fertility management but is also helpful in weed, insects and disease control. Legumes are essential in rotation practices for nitrogen supplement to the soil. The practice of mixed cropping increases the crop yield and avoids the chances of disease occurrence and pest infestation.

Advantages and Disadvantages of Organic Farming

These are the organic farming pros and cons to consider when looking at this practice.

List of the Pros of Organic Farming

Organic Farming does not use Genetically-modified Products

If you are choosing organic foods, then you are purchasing products which are not genetically modified using artificial methods. GMOs are not allowed within the industry. You will still experience products that are cross-bred naturally by the farmers to take advantage of specific growth characteristics, but this process is done through pollination or breeding instead of genetic splicing and dicing.

Organic Farming Helps to Support Healthier Soils

Farmers often use a pattern of crop rotation as a way to support a healthy balance of nutrients in their soil. A standard rotational process in the U.S. Midwest might involve planting corn for one year, then planting oats for the second year, before allowing the land to be used for pasture or hay crops for the next 2-6 years. This process creates higher content levels of organic matter and better aeration that can promote better productivity.

Better soils happen because there are no synthetic herbicides, pesticides, fertilizers, or chemicals applied to the crops grown on the farm either. Only natural enhancements of the soil are permitted if the land is going to be certified for organic use.

Organic Farming Produces Foods that are Better for you

There are numerous boosts to flavor profiles and nutritional content when organic farming is the preferred method in use. Peaches grown this way typically have a higher polyphenol content, which is why they taste better. Cows that are given room to graze in a pasture produce dairy products which contain higher levels of Vitamin E and Omega-3s to support a stronger antioxidant profile for consumers. Conventional farming might produce a greater quantity of food items, but organic farming improves the overall quality of the food so that consumers don't have to eat as much.

Organic Farming Support Pollinating Insects

Farmers that use organic methods are not using synthetic agents like glyphosate or neonicotinoids that create a high risk for harm to the pollinating insects that support crop systems all over the world. These substances were shown by Harvard researchers to be one of the primary causes of Colony Collapse Disorder. Up to 40% of honeybee colonies died between 2014-2015 and found that neonicotinoids, which are the most widely used insecticides for commercial growing, were responsible for six times more loss when compared to Nosema, which is a parasitic fungus.

Organic Farming Creates a Healthier Work Environment

Local communities, employees, and even the farmers themselves are not exposed to synthetic agricultural chemicals when following organic process. When too much exposure occurs, there can be issues with toxicity that occur. People who come into contact with pesticides regularly as part of their job duties suffer a higher risk of neurological disease when compared to the general population. Farmers can suffer from a variety of bothersome symptoms as well, ranging from memory loss to headaches to chronic fatigue.

Organic Farming Promotes a Greater Resistance to Pests and Disease

Healthier plants grow when the soil foundations that support their roots have a natural profile of disease resistance. This process encourages the plants to become naturally resistant to disease and pests because they have a stronger immune profile. Their defense mechanisms work to repel invaders when there is an appropriate pH level and other optimal conditions present, such as sunlight and water. It creates a thicker plant cell wall that creates healthier plant growth overall.

Organic Farming uses Natural Fertilizers to Encourage better Yields

Organic farmers do not plant their seeds and then let nature run its course. They use a variety of soil fertility methods which support the growth infrastructure of their crops. Techniques like the use of green manure, worm farming, compost application, and cover crops help to reinforce the stability of the soil while managing pests, weeds, and other potential hazards that workers encounter during the growing season. Each method helps to maintain the long-term productivity of the fields while encouraging higher levels of biodiversity at the local level.

Organic Farming Offers Profitable Niche Crop Opportunities

One of the most significant advantages of organic farming is the opportunity to begin diversifying the products which are cultivated each year. Conventional farms rely on cash crops as a way to earn profits to continue their operations. That is why soybeans and corn become the primary crops grown at these facilities. Organic farmers have an opportunity to grow several different varieties of items, including heirloom produce that can be sold throughout the year at a higher overall price.

Heirloom crops offer an exceptional taste profile, the potential for better nutrition, and the opportunities to save seeds each year for additional replanting. They are less uniform during ripening to create better product availability. Their seeds are usually less expensive than hybrids as well.

Organic Farming is an Eco-friendly Method of Growing to use

Organic farming offers benefits to our climate to consider as well. The processes involved with this method help to store carbon in our soils. It reduces the energy requirements necessary to produce a crop because physical labor is used more often than mechanical tools. There is a reduction in the use of petroleum-based products when taking the organic approach, which means fewer greenhouse gas emissions to consider. It will even support the natural ecosystems that store carbon as well, such as our prairies and forests.

Organic Farming can Help Future Generations Find Success

Organic farming methods focus on creating a restorative process for our soils that makes it possible to continue using them indefinitely for crop production. Instead of focusing on significant short-term profits that potentially destroy the land and the surrounding natural resources, this process encourages us to live and work within our means. It seeks to increase the natural capital values found on our planet instead of reducing them. That's why organic methods, when they are correctly implemented, can help to create long-term sustainable food chains that will support a growing population.

List of the Cons of Organic Farming

Organic Farming Operations are Rarely Subsidized

This disadvantage primarily applies to the United States. There are programs available, such as the Environmental Quality Incentives Program, that will pay producers to transition from conventional farming methods to organic work. Some programs will assist farmers in the costs of certifying their land as being organic. There are crop insurance subsidies available as well. What you won't find are the direct payments made to inflate the pricing schemes and artificial methods to enhance yields that can make going organic financially challenging for some.

Organic Farming Lacks a Supportive Infrastructure in the United States

European farmers get to avoid this disadvantage of organic farming as well. There is a lack of special infrastructure in the United States that supports the natural methods used in this agricultural style. That means organic farmers use the same industrial transportation methods to get food on the tables of consumers as the cash crop farmers use for their yields. That means the same harmful practices that are in place can still create environmental damage even if soil-friendly methods are used to grow the crops initially.

Organic Farming Still uses Fungicides and Pesticides

The keyword to consider when looking at organic farming processes is this: "synthetic." There are still several fungicides and pesticides which are available to use when following organic farming processes. The only requirement is that the product must come from a natural source instead of a synthetic one. These items are still potentially harmful because they require repetitive applications, promoting soil storage of the compounds that may exceed safe concentration levels in some areas.

Organic Farming does not always Account for Previous Practices

Transitioning from conventional farming to an organic approach is something that benefits the world's food chain thanks to all of the advantages involved. Unfortunately, the issue of synthetic chemicals staying in the soil can exist for farmers who switch their practices for decades because of their concentration levels. Farms can still sell products that are labeled as "organic" in this situation because their current methods follow the published guidelines. The food might still contain trace amounts of synthetic items despite the labeling and practices involved because of previous production methods on the same land.

Organic Farming Requires more Work than Conventional Methods

Organic farming requires a lot of physical work to create a successful experience. It needs workers to physically control the weeds and apply cultivation techniques. There are ways to reduce this disadvantage by focusing on biointensive farming or permaculture, but it also requires more of a personal touch than what conventional cash-farming methods require. Even if you take the time to experiment with different production methods, it can take years of trial-and-error to find the best combination of methods that maximize profits.

Organic Farming Requires Specific Knowledge to be Successful

Organic farmers must have an understanding of their land's soil ecology. They must know how to develop natural systems that work with their climate and crops. Farmers must invest time in learning about industry innovations or finding alternative solutions because the support systems that help conventional farms are not always available. This process relies on the experience of workers as they monitor crops during the crucial periods of growth to ensure the plants grow in healthy ways. Without this knowledge and the ambition to apply it, this method cannot be successful.

Organic Farming Faces Several Marketing Challenges to Consider

Conventional farmers have a defined market which allows them to sell their produce and farm products. They can ship to grocery stores, access a lucrative export market, and still reach local consumers. Organic products face a different challenge. Even if there are opportunities for farming co-ops, grocery store inclusion, and local markets that operate on specific days, this industry does not have the same levels of access for product sales in most communities. That makes it difficult for the farmers to maximize their profit potential when selling their products.

Organic Farmers must go through a Significant Certification Process

Did you know that there are different levels of "organic" in the United States? Each product goes through a different certification procedure to have this labeling opportunity available. Items that are 100% organic receive a specific label which indicates every process follows the industry standards. If something is just "organic," then 95% or more of the ingredients are organic, but there can be USDA-approved chemical additives added to the item before it reaches the market. If something is "made from organic ingredients," then only specific items follow industry processes.

The pros and cons of organic farming seek to find a balance between the better nutrition that is available from this method with the access challenges that workers face when trying to sell their product to local consumers. Even with the potential for subsidies and grants to offset some of the costs of becoming an organic producer, the disadvantages found in this industry can drive some farmers away.

Organic Food

Organic food is the fresh or processed food produced by organic farming methods. Organic food is grown without the use of synthetic chemicals, such as human-made pesticides and fertilizers, and does not contain genetically modified organisms (GMOs). Organic foods include fresh produce, meats, and dairy products as well as processed foods such as crackers, drinks, and frozen meals. The market for organic food has grown significantly since the late 20th century, becoming a multi-billion dollar industry with distinct production, processing, distribution, and retail systems.

fresh or processed food produced by organic farming methods. Organic food is grown without the use of synthetic chemicals, such as human-made pesticides and fertilizers, and does not contain genetically modified organisms (GMOs). Organic foods include fresh produce, meats, and dairy products as well as processed foods such as crackers, drinks, and frozen meals. The market for organic food has grown significantly since the late 20th century, becoming a multibillion dollar industry with distinct production, processing, distribution, and retail systems.

Policy

Although organic food production began as an alternative farming method outside the mainstream, it eventually became divided between two distinct paths: (1) small-scale farms that may not be formally certified organic and thus depend on informed consumers who seek out local, fresh, organically grown foods; and (2) large-scale certified organic food (fresh and processed) that is typically transported large distances and is distributed through typical grocery store chains. If consumers know their local farmer and trust the farmer's production methods, they may not demand a certification label. On the other hand, organic food produced far away and shipped is more likely to require a certification label to promote consumer trust and to prevent fraud, which exemplifies how national certification regulations are most beneficial.

Farmers' market Display of organic produce at an outdoor farmers' market

A regulatory framework is most important when consumers and farmers are geographically separated, and such a framework is likely to cater to larger-scale producers who participate in a more industrial system. This regulatory approach does not necessarily match consumers' assumptions about organic food production, which typically include images of small family farms and the humane treatment of animals. In general, regulations surrounding organic food do not address more complex social concerns about family farms, farmworker wages, or farm size, and organic policy in some places does little to address animal welfare.

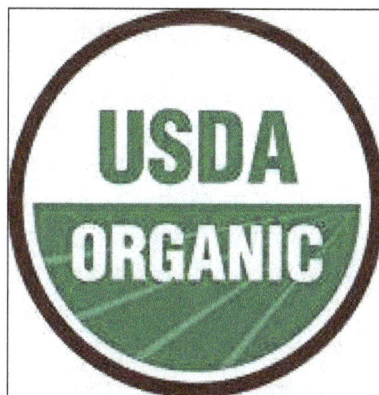

Certified organic The organic seal of the United States Department of Agriculture (USDA)

Organic food policies were created largely to provide a certification system with specific rules regarding production methods, and only products that follow the guidelines are allowed to use the certified organic labels. In the United States, the Organic Foods Production Act of 1990 began the process of establishing enforceable rules to mandate how agricultural products are grown, sold, and labeled. The regulations concerning organic food and organic products are based on a National List of Allowed and Prohibited Substances, which is a critical aspect of certified organic farming methods. The United States Department of Agriculture (USDA) regulates organic production through its National Organic Program (NOP), which serves to facilitate national and international marketing and sales of organically produced food and to assure consumers that USDA certified organic products meet uniform standards. To this end, NOP established three specific labels for consumers on organic food products: "100% organic," "organic," or "made with organic," which signify that a product's ingredients are 100 percent, at least 95 percent, or 70 percent organic, respectively. Noncertified products cannot use the USDA organic seal, and violators face significant fines and penalties.

Organic regulations vary by country, some of the most comprehensive rules being seen in Europe. Objectives of organic farming in the European Union (EU) include respecting nature's biological systems and establishing a sustainable management system; using water, soil, and air responsibly; and adhering to animal welfare standards that meet species-specific behavioral needs. In addition, principles of organic production in the EU are based on designing and managing farms to promote ecological systems and on using natural resources within the farming system. These policy goals go far beyond a defined listing of prohibited materials in organic production.

Environment

The overall impacts of organic agriculture are beneficial to the environment. Certified organic production methods prohibit the use of synthetic fertilizers and pesticides, thus reducing chemical runoff and the pollution of soils and watersheds. Smaller-scale organic farming often is associated with significant environmental benefits, owing to the use of on-farm inputs, such as fertilizers derived from compost created on-site. By comparison, large-scale organic farms often require inputs generated off-site and may not employ integrated farming methods. These operations may buy specific allowable inputs, such as fish emulsion or blood meal to use as fertilizer rather than working within the farm to increase soil fertility. While this decrease in synthetic chemical use benefits the environment compared with industrial agriculture, these methods may not promote long-term sustainability, since off-farm inputs usually require greater fossil fuel use than on-farm inputs.

Society

Social concerns related to organic food include higher costs to consumers and geographic variations in demand. Organic food usually is more expensive for consumers than conventionally produced food because of its more labour-intensive methods, the costs of certification, and the decreased reliance on chemicals to prop up crop yields. This often translates into unequal access to organic food. Research indicates that greater wealth and education levels are correlated with organic food purchases. Further, there are trends in some lower-income countries to produce certified organic crops solely for export to wealthier countries. This sometimes generates a situation in which the farmers themselves cannot afford to buy the organic foods they are producing. While this strategy may bring economic gain in the short term, it is a concern when farmers are forced out of producing food crops that feed their local communities, thus increasing food insecurity.

Certified organic agriculture has also become a big business in many places, with larger farming operations playing a key role in national and global certified organic food markets. Given economies of scale, big food-processing companies often buy from a single farming operation that produces organic crops on thousands of acres, rather than from many smaller farms that each grow on smaller acreages, a practice that effectively limits the participation of smaller farmers in these markets. There also is disparity among farmers, since the organic certification process can be prohibitively expensive to some smaller-scale farmers. Although certification subsidies exist in some places, such farmers often opt to sell directly to consumers at farmers' markets, for example, and may decide to forgo organic certification altogether.

Overall, organic food has grown in popularity, as consumers have increasingly sought and purchased foods that they perceive as being healthier and grown in ways that benefit the environment. Indeed, consumers typically buy organic food in order to reduce their exposure to pesticide residues and GMOs. Further, some research shows that organically produced crops have higher nutritional content than comparable nonorganic crops, and some people find organic foods to be tastier. The question remains, however, whether organic food shipped in from across the globe is truly a sustainable method of food production. Certainly organically produced food from a local farmer who employs an integrated whole-farm approach is fairly environmentally sustainable, though the economic sustainability of such an endeavour can be challenging. Although humans must decrease their reliance on fossil fuels to combat climate change, many organic policies do little to address the issue of sustainability, focusing instead on the strict list of prohibited substances, rather than a comprehensive long-term view of farming and food.

Organic crops must be grown without the use of synthetic pesticides, bioengineered genes (GMOs), petroleum-based fertilizers, and sewage sludge-based fertilizers.

Organic livestock raised for meat, eggs, and dairy products must have access to the outdoors and be given organic feed. They may not be given antibiotics, growth hormones, or any animal by-products.

Table: Organic vs. Non-Organic

Organic vs. Non-Organic	
Organic produce:	Conventionally-grown produce:
Grown with natural fertilizers (manure, compost).	Grown with synthetic or chemical fertilizers.
Weeds are controlled naturally (crop rotation, hand weeding, mulching, and tilling).	Weeds are controlled with chemical herbicides.
Pests are controlled using natural methods (birds, insects, traps) and naturally-derived pesticides.	Pests are controlled with synthetic pesticides
Organic meat, dairy, eggs:	Conventionally-raised meat, dairy, eggs
Livestock are given all organic, hormone- and GMO-free feed.	Livestock are given growth hormones for faster growth, as well as non-organic, GMO feed.
Disease is prevented with natural methods such as clean housing, rotational grazing, and healthy diet.	Antibiotics and medications are used to prevent livestock disease.
Livestock must have access to the outdoors.	Livestock may or may not have access to the outdoors.

Benefits of Organic Food

How your food is grown or raised can have a major impact on your mental and emotional health as well as the environment. Organic foods often have more beneficial nutrients, such as antioxidants, than their conventionally-grown counterparts and people with allergies to foods, chemicals, or preservatives often find their symptoms lessen or go away when they eat only organic foods.

Organic produce contains fewer pesticides. Chemicals such as fungicides, herbicides, and insecticides are widely used in conventional agriculture and residues remain on (and in) the food we eat.

Organic food is often fresher because it doesn't contain preservatives that make it last longer. Organic produce is often (but not always, so watch where it is from) produced on smaller farms near where it is sold.

Organic farming is better for the environment. Organic farming practices reduce pollution, conserve water, reduce soil erosion, increase soil fertility, and use less energy. Farming without pesticides is also better for nearby birds and animals as well as people who live close to farms.

Organically raised animals are not given antibiotics, growth hormones, or fed animal by-products. Feeding livestock animal by-products increases the risk of mad cow disease (BSE) and the use of antibiotics can create antibiotic-resistant strains of bacteria. Organically-raised animals are given more space to move around and access to the outdoors, which help to keep them healthy.

Organic meat and milk are richer in certain nutrients. Results of a 2016 European study show that levels of certain nutrients, including omega-3 fatty acids, were up to 50 percent higher in organic meat and milk than in conventionally raised versions.

Organic food is GMO-free. Genetically Modified Organisms (GMOs) or genetically engineered (GE) foods are plants whose DNA has been altered in ways that cannot occur in nature or in traditional crossbreeding, most commonly in order to be resistant to pesticides or produce an insecticide.

Organic Food vs. Locally-grown Food

Unlike organic standards, there is no specific definition for "local food". It could be grown in your local community, your state, your region, or your country. During large portions of the year it is usually possible to find food grown close to home at places such as a farmer's market.

Benefits of Locally Grown Food

- Financial: Money stays within the local economy. More money goes directly to the farmer, instead of to things like marketing and distribution.

- Transportation: In the U.S., for example, the average distance a meal travels from the farm to the dinner plate is over 1,500 miles. Produce must be picked while still unripe and then gassed to "ripen" it after transport. Or the food is highly processed in factories using preservatives, irradiation, and other means to keep it stable for transport.

- Freshness: Local food is harvested when ripe and thus fresher and full of flavor.

Small local farmers often use organic methods but sometimes cannot afford to become certified organic. Visit a farmer's market and talk with the farmers to find out what methods they use.

Understanding GMOs

The ongoing debate about the effects of GMOs on health and the environment is a controversial one. In most cases, GMOs are engineered to make food crops resistant to herbicides and to produce an insecticide. For example, much of the sweet corn consumed in the U.S. is genetically engineered to be resistant to the herbicide Roundup and to produce its own insecticide, Bt Toxin.

GMOs are also commonly found in U.S. crops such as soybeans, alfalfa, squash, zucchini, papaya, and canola, and are present in many breakfast cereals and much of the processed food that we eat. If the ingredients on a package include corn syrup or soy lecithin, chances are it contains GMOs.

GMOs and Pesticides

The use of toxic herbicides like Roundup (glyphosate) has increased 15 times since GMOs were introduced. While the World Health Organization announced that glyphosate is "probably carcinogenic to humans," there is still some controversy over the level of health risks posed by the use of pesticides.

Safety of GMOs

While the U.S. Food and Drug Administration (FDA) and the biotech companies that engineer GMOs insist they are safe, many food safety advocates point out that no long term studies have ever been conducted to confirm the safety of GMO use, while some animal studies have indicated that consuming GMOs may cause internal organ damage, slowed brain growth, and thickening of the digestive tract.

GMOs have been linked to increased food allergens and gastro-intestinal problems in humans. While many people think that altering the DNA of a plant or animal can increase the risk of cancer, the research has so far proven inconclusive.

Difference between Organic and Pesticide-free

One of the primary benefits of eating organic is lower levels of pesticides. However, despite popular belief, organic farms do use pesticides. The difference is that they only use naturally-derived pesticides, rather than the synthetic pesticides used on conventional commercial farms. Natural pesticides are believed to be less toxic, however, some have been found to have health risks. That said, your exposure to harmful pesticides will be lower when eating organic.

Possible Risks of Pesticides

Most of us have an accumulated build-up of pesticide exposure in our bodies due to numerous years of exposure. This chemical "body burden" as it is medically known could lead to health issues such as headaches, birth defects, and added strain on weakened immune systems.

Some studies have indicated that the use of pesticides even at low doses can increase the risk of certain cancers, such as leukemia, lymphoma, brain tumors, breast cancer and prostate cancer.

Children and fetuses are most vulnerable to pesticide exposure because their immune systems, bodies, and brains are still developing. Exposure at an early age may cause developmental delays, behavioral disorders, autism, immune system harm, and motor dysfunction.

Pregnant women are more vulnerable due to the added stress pesticides put on their already taxed organs. Plus, pesticides can be passed from mother to child in the womb, as well as through breast milk.

The widespread use of pesticides has also led to the emergence of "super weeds" and "super bugs," which can only be killed with extremely toxic poisons like 2,4-Dichlorophenoxyacetic acid (a major ingredient in Agent Orange).

Does Washing and Peeling Produce Get Rid of Pesticides?

Rinsing reduces but does not eliminate pesticides. Peeling sometimes helps, but valuable nutrients often go down the drain with the skin. The best approach: eat a varied diet, wash and scrub all produce thoroughly, and buy organic when possible.

Expensive

Organic food is often more expensive than conventionally-grown food. But if you set some priorities, it may be possible to purchase organic food and stay within your food budget.

Know your Produce Pesticide Levels

Some types of conventionally-grown produce are much higher in pesticides than others, and should be avoided. Others are low enough that buying non-organic is relatively safe. The Environmental Working Group, a non-profit organization that analyzes the results of government pesticide testing in the U.S., offers an annually-updated list that can help guide your choices.

Fruits and Vegetables where the Organic Label Matters Most

According to the Environmental Working Group, a nonprofit organization that analyzes the results of government pesticide testing in the U.S., the following fruits and vegetables have the highest pesticide levels so are best to buy organic:

- Apples,
- Sweet Bell Peppers,
- Cucumbers,
- Celery,
- Potatoes,
- Grapes,
- Cherry Tomatoes,
- Kale/Collard Greens,
- Summer Squash,
- Nectarines (imported),
- Peaches,

- Spinach,

- Strawberries,

- Hot Peppers.

Fruits and Vegetables you don't Need to Buy Organic

Known as the "Clean 15", these conventionally-grown fruits and vegetables are generally low in pesticides.

- Asparagus,

- Avocado,

- Mushrooms,

- Cabbage,

- Sweet Corn,

- Eggplant,

- Kiwi,

- Mango,

- Onion,

- Papaya,

- Pineapple,

- Sweet Peas (frozen),

- Sweet Potatoes,

- Grapefruit,

- Cantaloupe.

Organic Meat, Eggs and Dairy

While prominent organizations such as the American Heart Association maintain that eating saturated fat from any source increases the risk of heart disease, other nutrition experts maintain that eating organic grass-fed meat and organic dairy products doesn't carry the same risks. It's not the saturated fat that's the problem, they say, but the unnatural diet of an industrially-raised animal that includes corn, hormones, and medication.

Organic Certification of Vegetable Operations

All farms and ranches, including vegetable growers, who sell over $5000 per year of organic products, must be certified in order to sell their products as organic. Land used for the production of organic vegetables must not have had prohibited fertilizers, pesticides, GMOs, or other prohibited substances applied for at least 36 months prior to the first harvest of an organic crop. Farmers who sell under $5000 per year of organic produce must still follow all provisions of the USDA organic regulations, but are not required to be certified as organic. Non-certified organic operations who sell less than $5000 per year can only sell their products directly to retailers and consumers. Their products cannot be sold as organic feed or as organic ingredients that will be further processed and subsequently labeled as organic. Farms can be certified as organic as a whole farm or on a field-by-field basis.

Transitioning to Organic Production

When you begin transitioning your vegetable operation to organic production, choose a USDA-accredited certification agency, and request an Organic System Plan (OSP) questionnaire or application packet. During the final 12 months of the 36 month transition period, you should begin the application process, so that your operation can become certified as soon as your transition is complete. For most growers, the application, inspection, and approval process takes between three to six months, so don't wait until the last minute to seek certification.

It is important to document the last date when a prohibited substance was applied, in order to demonstrate to the certification agency that the field has been free of prohibited substance applications for 36 months and is eligible for organic certification. This is especially important for vegetable growers who produce crops such as lettuce or spinach, which might be harvested early in a given year. In such cases, if you can document the last application of a prohibited substance as being in May of transition year one, for example, then you can harvest organic crops from that field in June of transition year three, since the 36-month transition period has elapsed.

You will need to develop and submit to your certifier a field history sheet, showing the crops grown and inputs applied for at least the past 3 years, for all fields requested for certification. If the land has not been under your control for 36 months prior to the projected harvest of your first organic crop, you will need to obtain and submit a signed statement from the previous farm operator (owner or renter) providing information on the crops grown, inputs, and production practices during the transition period.

During transition, you should establish a soil-building crop rotation and develop effective fertility, pest, disease, and weed management strategies using preventive practices and natural fertility inputs such as compost, mulch, and cover crops. If needed, you may use non-synthetic (natural) biological, botanical, or mineral inputs, or, if these are not effective, synthetic substances that appear on the National List of Approved and Prohibited Substances, which is part of the NOP regulation.

Where prohibited substances are or will be applied to fields, roadsides, drainage ditches, railroad right-of-ways, or under utility lines that adjoin your organic production areas, you should establish buffer zones wide enough to prevent drift of prohibited substances onto the land you are transitioning. During the entire 36-month transition, as well as when you are certified organic, you must: discontinue all uses of prohibited substances, including fungicide-treated seeds, chemical fertilizers, and non-approved synthetic pesticides; implement conservation practices; and set up an appropriate record-keeping system, so that you can track all seeds, seedlings, inputs used, and crops harvested, stored, and sold.

Certified Organic Seed

You are required to use certified organic seed for crops you wish to have certified, unless you can document that the seed you need to plant is not commercially available from organic sources in the form, quality, quantity, or equivalent variety that you need for your operation. Proof must be provided to your certifier that you made good-faith efforts to obtain organic seed. This proof can consist of written records documenting telephone calls, results from searches in seed catalogs, or letters from organic seed suppliers stating that certified organic seed was not available. High price is not an acceptable reason for not purchasing organic seed. If certified organic seed is documented as not available, untreated, conventionally-grown seed may be used. Genetically engineered seeds cannot be used. A "Non-GMO Affidavit" should be obtained from seed suppliers for all purchased non-organic seed that has a GMO equivalent. Cover crop seeds used for incorporation as a green manure are also required to be certified organic, unless you provide evidence that organic seed is commercially unavailable.

Organic Seedlings and Transplants

Annual seedlings used for organic crops must be certified organic. "Annual seedling" is defined by the NOP as, "a plant grown from seed that will complete its life cycle or produce a harvestable yield within the same crop year or season in which it is planted." This means that organic seedlings must be used for crops such as tomatoes, peppers, many brassicas, and other crops that are not direct-seeded. The only exception is for organic crops destroyed by natural disasters, in which case the certifier, if approved by the NOP based on a declaration of natural disaster, can grant a temporary variance for the crop to be re-planted using conventionally-grown seedlings. In all other instances, use of nonorganic annual seedlings will jeopardize your organic certification for that year, as well as possibly for future years.

Planting stock, used for the production of potatoes, or for perennial crops such as raspberries, strawberries, or apples, falls under a different requirement. "Planting stock" is defined by the NOP as, "any plant or plant tissue other than annual seedlings but including rhizomes, shoots, leaf or stem cuttings, roots, or tubers, used in plant production or propagation." First, the organic producers must attempt to source organic planting stock, just as you must attempt to source organic seeds. If organic planting stock is documented as commercially unavailable, conventionally-grown planting stock may be used to produce organic crops. The planting stock itself can only be sold as organic after it has been grown organically for one year. The crop harvested from the planting stock can be sold as organic.

Greenhouses

In order to produce organic seedlings, or to produce organic crops in high tunnels or other structures, greenhouses must use organic methods and approved inputs, and they need to be described in the operation's Organic System Plan, or certified on their own. The greenhouse operator, whether on-farm or off-farm, needs to list all organic and non-organic crops grown, and list all fertility, pest, and disease inputs used or planned for use in the greenhouse. Natural materials, such as compost, sand, peat, vermiculite, perlite, and natural rock minerals are commonly used in soil mixes or to fertilize in-ground organic greenhouse production. Carefully review the ingredients in purchased soil mixes; synthetic wetting agents, fumigants, or synthetic fertilizers are not allowed. Often, these ingredients do not appear on product labels, so check with the manufacturer or your certifier to make sure the product is allowed.

Natural botanical, biological, or mineral inputs may be used for pest and disease control. If needed, approved synthetic substances on the National List may also be used. It is always a good idea to check with your certifier before purchasing and applying any input, to make sure that it is allowed for organic production. Make sure to keep receipts and label information for all inputs used or planned for use.

Planting trays, pots, and irrigation lines can be cleaned and sanitized using hot water, alcohols, chlorine materials, hydrogen peroxide, ozone gas, peracetic acid, or soaps. Once again, check with your certifier to make sure that the substance you plan to use is approved. Both active ingredients and secondary ingredients must be approved for items that are in compounded materials.

If the greenhouse is also used for the production of non-organic plants, there must be clear separation between the organic and non-organic areas to prevent contamination, and clear identification and tracking of the organic vs. non-organic plants. If synthetic fertilizers are injected into the water system for the non-organic crops, the injection system must be disabled or a separate watering system used to assure that the prohibited fertilizers are not used on the organic seedlings or crops.

Treated Lumber

The NOP prohibits the use of lumber treated with arsenate compounds or other prohibited substances for new installations or replacement purposes, where the lumber will contact the crop, soil, or livestock. This means that treated wood should not be used for plant trays, trellises, or posts, where the treated wood will come in contact with organic crops or soil used to produce organic crops. Not all wood preservatives listed in the ATTRA publication are approved for organic use. Check with your certifier before using any treated wood. If treated lumber was in place before organic certification, then it may remain, but be aware that some of the toxins in the lumber may be taken up by plants growing nearby.

Crop Rotation

NOP section requires all organic producers, including organic vegetable growers, to implement crop rotations that include, but are not limited to "sod, cover crops, green manure crops, and catch crops" to: a) maintain or improve soil organic matter content; b) provide for pest management; c) manage deficient or excess nutrients, and d) provide erosion control.

The NOP defines "crop rotation" as, "the practice of alternating the annual crops grown on a specific field in a planned pattern or sequence in successive crop years so that crops of the same species or family are not grown repeatedly without interruption on the same field. Perennial cropping systems employ means such as alley cropping, intercropping, and hedgerows to introduce biological diversity in lieu of crop rotation."

All organic vegetable producers must implement crop rotations, which meet the objectives listed above. Using crop rotations to break insect, disease and weed cycles and improve soil fertility can aid the organic vegetable grower in producing high quality crops.

Manure and Compost

The NOP regulation has strict requirements on the use of manure and compost in organic production systems. All animal manure must be composted if applied to vegetable crops destined for human consumption, or else certain restrictions apply. If the manure is fresh, or has not gone through a complete composting process, it must be incorporated into the soil at least 120 days before a vegetable crop will be harvested, if the edible portion of the crop comes into contact with the soil or soil particles. In regions where cold limits the growing season, all raw manure should be incorporated in the field during the fall prior to vegetable crop planting, in order to comply with the 120-day waiting period. If the edible portion of the crop does not come into contact with the soil (e.g. sweet corn), raw manure may be incorporated into the soil at least 90 days before harvest.

Compost may be applied at any time. "Compost" is defined by the NOP as, "the product of a managed process through which microorganisms break down plant and animal materials into more available forms suitable for application to the soil. Compost must be produced through a process that combines plant and animal materials with an initial C:N ratio of between 25:1 and 40:1. Producers using an in-vessel or static aerated pile system must maintain the composting materials at a temperature between 131 F and 170 F for 3 days. Producers using a windrow system must maintain the composting materials at a temperature between 131 F and 170 F for 15 days, during which time, the materials must be turned a minimum of five times."

Organic producers making their own compost must keep records of their composting operation to demonstrate that the compost was produced according to the definition cited above. If the compost is purchased, the grower should ask for documentation from the supplier showing that the compost meets NOP requirements. Keep this documentation, along with purchase receipts, with your other records. If the compost is 100% plant-based, without any animal excrement or by-products, there is no requirement for heating or turning.

Heat-treated, processed manure may be used as a supplement to a soil building program, without a specific interval between application and harvest. Producers are expected to comply with all applicable requirements of the NOP regulation with respect to soil quality, including ensuring the soil is enhanced and maintained through proper stewardship.

According to the NOP's July 17, 2007, ruling, "processed manure products must be treated so that all portions of the product, without causing combustion, reach a minimum temperature of either 150 °F (66 °C) for at least one hour or 165 °F (74 °C), and are dried to a maximum moisture level of 12%; or an equivalent heating and drying process could be used. In determining the acceptability of an equivalent process, processed manure products should not contain more than 1×10^3 (1,000) MPN (Most Probable Number) fecal coliform per gram of processed manure sampled and not contain more than 3 MPN Salmonella per 4 gram sample of processed manure."

As always, organic vegetable growers should get label information and check with their certifiers, before using purchased compost or processed manure products.

Prohibited and Approved Inputs

Prohibited substances are typically synthetic substances that are not allowed under the NOP, although there are a few natural substances that fall into this category as well. Prohibited substances include chemical fertilizers and synthetic herbicides, fungicides, and insecticides, as well as genetically engineered organisms, which are referred to as "excluded methods" by the NOP. Prohibited substances include items such as seeds treated with Captan, Thiram or with genetically modified rhizobial bacteria. All synthetic materials are prohibited for use, unless they have been specifically approved by the NOP and appear on the National List. All natural products are allowed, unless they are specifically listed as prohibited on the same list.

The organic regulation mandates that a hierarchy be followed for pest, disease, and weed control. You must start with cultural controls (i.e. disease-resistant varieties or the timing of planting), mechanical controls (i.e. the use of row covers or flaming weeding), or biological (i.e. the use of beneficial insects). If these methods are documented as ineffective, then natural products can be used. If natural products are not effective, then approved synthetic products can be used.

For pest control products, the active ingredients and the inert ingredients must be allowed for organic production. The acceptability of brand name products should be verified with your certification agent.

Mulches

Mulching is an approved weed control option, with natural materials such as straw, tree leaves, or grass clippings being allowed, so long as the mulch does not pose a risk of contamination with herbicide residues (such as lawn clippings from chemically-treated lawns). Plastic mulches are allowed, but must be removed at the end of the growing or harvest season.

Conservation and Biodiversity

The NOP defines "organic production" as a "production system that is managed in accordance with the Act and regulations in this part to respond to site-specific conditions by integrating cultural, biological, and mechanical practices that foster cycling of resources, promote ecological balance, and conserve biodiversity." Promotion of ecological balance and conservation of biodiversity are inherent to organic production.

The NOP requires that organic producers must maintain or improve the natural resources of their operations, including soil and water quality, and minimize soil erosion. Organic vegetable growers comply with these requirements by implementing conservation practices, such as crop rotations, cover crops, grass waterways, and contour strips. Many grow annual and perennial flowering plants, which provide food and habitat for pollinators and other beneficial organisms. Some also erect bird and bat houses to enhance biodiversity, while improving pest control for crops.

Harvest and Storage

During and after harvest, certified organic produce must be kept separate from non-organic produce. There can be no commingling of organic and non-organic products or contamination through contact with prohibited substances. Equipment that is used to harvest conventional crops as well as organic crops must be thoroughly cleaned prior to organic harvest. The grower must document equipment cleaning activities on a "Cleaning Affidavit" or such record. Wagons made from treated wood should be covered with a tarp to avoid direct contact with organic produce. Treated wood bins or waxed boxes that previously held nonorganic produce should not be used for organic crops.

Storage areas used for organic products must be separate and labeled as such, especially if organic and non-organic products are stored in the same facility. If a walk-in cooler is used for organic and non-organic products, the organic products should never be stored below non-organic products, as contaminated water could easily drip onto the organic products.

Issues in Organic Agriculture

Ecological Dimensions

The flow of energy (that involves biological and non-biological agents) drives the carbon, oxygen, nitrogen and phosphorus cycles. Nutrients are pumped through the system by the action of photosynthesis and are again made available for recycling by the action of decomposers. Nutrients are constantly being removed or added; adding more natural substances or synthetic materials than the ecosystem is able to handle upsets bio-geochemical cycles.

For example, the nitrogen cycle is characterized by fixation of atmospheric nitrogen by nitrogen-fixing plants, largely legumes (i.e. symbiotic bacteria living in association with leguminous), root-noduled non-leguminous plants, free-living aerobic bacteria, and blue-green algae. In agricultural ecosystems, the nodulated legumes of approximately 200 species are the pre-eminent nitrogen fixers. In non-agricultural systems, some 12,000 species are responsible for nitrogen fixation.

Environmental Services that are Vital to Agriculture Include

- Soil forming and conditioning: A substantial amount of invertebrates (earthworms, millipedes, termites, mites, nematodes, etc.) play a role in the development of upper soil layers through decomposition of plant litter, making organic matter more readily available, and creating structural conditions that allow oxygen, food and water to circulate.

For example, the amount of soil worked over by earthworms is tremendous: 4-36 tons of soil passes through alimentary tracts of the total earthworm population living on an acre in a year! Termites are the only larger soil inhabitants that are able to break down the cellulose of wood. Termites play a major role in tropical soils where there are also soil churners; they move as much as 5 000 tons of soil per acre in constructing their complex mounds (allowing better rain penetration in soil).

- Waste disposal: A succession of micro-organisms occurs in the detritus, involving namely bacteria and fungi as well as detritus-feeding invertebrates, until organic material is finally reduced to elemental nutrients. Ecosystems recycle, detoxify and purify themselves, provided that their carrying capacity is not exceeded by excessive amounts of waste and by the introduction of persistent (synthetic) contaminants.

For example, the nutrient-filtering function of mangroves can be compared to that of oxidation ponds of conventional wastewater treatment plants:

- Pest control: Predation is not just the transfer of energy whereby one organism feeds on another organism but also complex interactions among predator-prey populations. If a portion of the prey is not available because of environmental discontinuities (a typical case in agriculture), the self-regulating balance will be dampened. Interspecific competition keeps more pests in check than we ever could by using pesticides.

- Biodiversity: An ecosystem stability (or instability) depends on the results of the competition between different species for food and space. Predation ameliorates the intensity of competition for space and increases species diversity. The nature of inter-specific competition and its effects on the species involved is one of the least known and most controversial areas of ecology.

- Beneficial associations: Symbiosis of plant roots with mycorrhizal fungi plays a most important role in temperate and tropical forests in absorbing nutrients, transferring energy and reducing pathogen invasions. Parasitism is used in the biological control of insects. Other symbiotic combinations include animal/fish/tree species (e.g. agroforestry, varietal diversification).

- Pollination: 220,000 out of 240,000 species of flowering plants are pollinated by insects.

- Carbon sequestration: The capacity of biomass in sequestrating carbon is receiving an increased attention with the aim of reducing (in the long term) climate change. Where no tillage is practised, soil contributes to retaining carbon. As organic agriculture favours minimum tillage (for better retention of water, nutrients, and biodiversity), the carbon retention potential of soils is becoming an important issue.

- Habitat: Although by definition, habitats provide shelter and food, many ecosystems have functions often discounted. For example, hedgerows around a field provide habitat for over-wintering of beneficial arthropods.

Natural Resources

Land

Organic farmers build on environmental services by using several techniques to prevent erosion (e.g. land consolidation and levelling) and enhance soil fertility. Soil structure is improved through nutrient mining (by deep rooting crops), improvement of nutrient availability with mycorrhizal and optimal nutrient recycling, specific crop rotation and manuring strategies (e.g. manure, compost, crop residues, legumes and green manures), and other natural fertilizers (e.g., rock phosphate, seaweed, wood ash). Minimum tillage avoids soil compaction. Integrating trees and shrubs conserves soil and water and provides a defence against unfavourable weather conditions such as winds, droughts, and floods.

Water

Maintaining water quality requires good knowledge of ecological processes (e.g. consolidating or developing vegetation that help water self-purification) and applying farm practices that avoid pollution and use little water (e.g. minimal use of irrigation, prevention of water evaporation losses). Due to the change in soil structure and organic matter content under organic management, water efficiency is likely to be high on organic farms. Efficient harvesting and use of water also controls salinization and waterlogging.

Genetic Material

Preservation of agro-biodiversity is indirectly addressed by organic agriculture since this form of agriculture relies mostly on endemic biodiversity that is resilient to local ecological stress (e.g. drought). The emphasis of seeds and breeds used in organic agriculture is on local suitability with respect to disease resistance and adaptability to local climate. The availability of suitable genetic material (e.g. GMO-free), its selection, rearing and distribution is often a constraint.

Agronomic Dimensions

Fertilizers and Pesticides

Decreasing the use of synthetic fertilizers and pesticides goes together with increasing other inputs. These inputs can be bought or produced on the farm (such as manure), others come in the form of knowledge about actions to be taken (e.g. timing of planting or best rotational combinations. The change in the combination of inputs may change the effectiveness of certain processes, which influence farm output, such as the cycles of water, nutrients, energy and knowledge (inter-generational). Crop protection relies on natural pest controls (e.g. insect pheromones, plants with pest control properties) or by enhancing self-regulation.

For example, a primary pest may be avoided by planting at a time when the insect cannot complete its life cycle, even though that results in a certain decrease in yield due to non-optimal conditions in other aspects such as heat; a secondary pest could stop after abandoning the use of pesticides and natural predators return.

Crop Rotation

Crop rotation is required under organic certification programmes and is considered to be the cornerstone of organic management. Crop rotation is a valuable tool for weed control, maintenance of soil structure and organic matter, recycling of plant nutrients, contribution to overall species and habitat diversity, preventing erosion (application of cover crops), green manuring, and pest and disease control. Inter-cropping, after-cropping, alleycropping and mixed cropping are spatial and temporal alternations chosen according to crop and soil-specific cycles and needs. Agricultural pests (fungi, insects, and weeds) are often specific to the host, and will multiply as long as the crop is there. Manipulation of crops between years (management by rotations) or within fields (strip-cropping) is therefore an important tool in the quest for management of pest problems. The lack of use of synthetic fertilizers and pesticides leads to restrictions in choice of crops. The loss in (present) income through a change in rotation is to some degree reflected in, and compensated by, the decrease in input costs. The success of an organic farm depends on the identification of end-uses and/or markets for all the crops in the rotation, as few farmers can afford to leave fields fallow. Wide adoption of organic agriculture is expected to change markets due to the likely prevalence of rotational crop products.

Energy

Energy in all its forms, including labour, is one of the most important agricultural inputs. Energy is required at all points of the food production chain: land preparation, planting, harvesting, transport, processing, irrigation, agroindustries and rural services (e.g. cooking, heating, and lighting). On organic farms, mechanization is often replaced by labour, especially for weeding and harvesting in highly diversified systems. Chemical fertilizers (that are produced using a high level of non-renewable energy) are substituted with organic matter, nitrogen fixing-crops or green manuring. Indirectly, using natural rather than synthetic fertilizers saves non-renewable energy and nitrogen leaching is minimized.

Disadvantages in discarding synthetic fertilizer must be considered as well: energy needs can escalate if thermal and mechanical weeding or intensive soil tillage is used and, in some cases, organic farmers burn to clear land which reduces fertility. Many resource-poor farmers do not have access to livestock manure, often an important fertility component. Sometimes sewage sludge is used, which may contain pathogens and other contaminants. Inappropriate management of energy inputs used in organic agriculture may be detrimental to the environment: over-manuring of nitrogen has several side effects such as leaching (hence affecting water quality), stimulation of plant diseases and discouraging biological N-fixation. Environmental standards are expected to reflect local conditions so that pollution is minimized. In fact, restrictions on the number of livestock or amount of manure to be used per unit of land are not exceptional.

Diversification

Organic agriculture requires a diversity of crops and livestock. Many indigenous food crops (e.g. yam, sorghum, millet, oil palm, cashew, mango) have been supplanted by monoproduction of cash crops. Food crops, pseudocereals (e.g. amaranth, buckwheat, chenopods), grain legumes (e.g. adzuki, faba, hyacinth beans) and other under-utilized plants, many of great value, can be reintroduced through crop rotations. This contributes to whole farm health, provides conservation of important genotypes, and creates habitats for beneficial species. Animal husbandry on organic farms focuses on manure production and the roughage and waste to food conversion capacity of animals.

Stability and Resilience

A system resilience is its ability to overcome perturbation and to recover its function to former levels once the perturbation is removed. Resilience is of particular value to farmers in terms of risk and productivity of the system. In organic agriculture in general, a diversity of crops is grown and different kinds of livestock are kept. This diversification means that the risk in variation in production is spread, as different crops react differently to climatic variation, or have different times of growing (both in the time of the year and in length of growing period). There is therefore less chance of a bumper year for all enterprises on organic farms (likely to coincide with relatively low prices). Also, there is less chance of low production for all crops and livestock simultaneously. This contributes to food security and stability of food supply.

Yields

Factors determining yields include plant varieties and knowledge on how to manipulate biological processes within agricultural systems. The management system is a major factor in the degree of yield and financial variability. Another factor, which makes a difference in yields, is the time and length of the growth period of a crop. Exogenous factors such as climate are important.

For example, due to slow mineralization of nitrogen under cool growing-conditions, crops on northern organic farms have a shortage of nitrogen early in the season. However, in countries where low soil temperature is not a limiting growth factor, as in many developing countries, this factor should not prove significant.

Although yields on organic farms generally fall within an acceptable range, yield comparisons depend on the systems involved. Organic agriculture yields are lower than yields of intensive external input agriculture. The latter, however, also results in considerable losses of nutrients and environmental quality as well as in food surpluses, which are exported with subsidies. Experiences of organic production in ecosystems with low-productivity potential demonstrate the potential to double or triple average yields. The results are of course due to very low initial yields on these lands but such conditions correspond to many countries of the developing world. If similar results were to be achieved in the less endowed regions of the world, present food deficits could be partly resolved. In any case, increased yields are more likely to be achieved if the departure point is a traditional system, even if degraded, rather than a modern system.

Total Farm Production

It is important to consider not only yields, but also whole farm production. The total production on the farm is the yield multiplied by the area in the different crops or that used for livestock (usually measured per unit of area). When measuring production, one also needs to be aware of the concept of net production, especially relevant in developing countries. This refers to the net production of specific inputs, such as the costs of nutrients.

For example, it is very easy to increase the yield of a cow by feeding her concentrates. The question is, however, whether it was worth the extra input. This can be determined by an assessment of the net returns to farming.

It should be realized that, during the conversion process, yields might be lower and investments higher than at a later stage when the organic farm has been established. The net returns to farming can therefore be lower in such a period than later.

Economic Dimensions

Farm Income

The suitability of an agricultural system depends on its profitability, if that concept includes all aspects which affect farmers' welfare.

For example, low return of a marketable crop as compared with another farming system may mean very little if inputs are also low, or if the farmer can harvest other products which can be grown simultaneously in the one system, but not in the other (such as fish with irrigated rice when no pesticides are used). In addition, relative incomes can change drastically with changing input or output prices.

In developed countries, the financial cost of inputs (excluding labour) on organic farms can be lower than on many non-organic farms, although the magnitude differs between enterprises and countries. The difference is generally greatest in those enterprises where inputs can be readily substituted by low-cost alternatives, as fertilizers by nitrogen fixing crops or green manure. For those inputs where substitutes are costly, such as labour cost for weeding, differences in expenditure on input between organic agriculture and other systems tend to be relatively low, or costs on organic farms can be higher. In situations where inputs are subsidized, as fertilizers and pesticides have been in a number of developing countries, the financial returns on organic farms may not be as attractive. Similarly, not counting the environmental and health costs of such inputs means that organic agriculture is under-valued.

The legal transition to organic agriculture takes two to three years during which products cannot be sold as organic. Initial loss of yields, extent to which inputs were used under the previous management system, and the state of ecosystem degradation are often constraints that can be easy to survive only if financial support is given to farmers. Hence, the degree of support during transition, and sometimes in the first years following the transition period, are important factors in farm economics. When organic agriculture performance on the environment is rewarded (e.g. through support to conversion), organic agriculture is as profitable as conventional agriculture.

Markets

Reliable market information, quantity and regularity of supply, and comparative advantages are key to tapping market opportunities. Output prices are subject to quantity of supply and consumer willingness to pay premiums for organic products (often at prices 20 percent higher than conventional products). Entering lucrative markets entails inspection, certification and labelling of produce, the cost of which being somewhat expensive. Factors such as farm size, volume of production, and efficiency (or availability) of certification organizations determine inspection costs. Often, small farmers cannot afford certification costs, and care should be taken for not marginalizing small producers.

Food Security

Food security is not necessarily achieved through food self-sufficiency. Consumers' demand for organically produced food and sometimes impressive premiums provide new export opportunities for farmers of the developing world. Returns from organic agriculture have the potential, under the right circumstances, to contribute to local food security by increasing family incomes. Organic agriculture can contribute to local food security in several ways. Organic farmers do not incur high initial expenses so less money is borrowed. Synthetic inputs, unaffordable to an increasing number of resource-poor farmers due to decreased subsidies and the need for foreign currency, are not used. Organic soil improvement may be the only economically sound system for resource-poor, small-scale farmers. This characteristic of the production process on organic farms means that organic farmers are less dependent on external inputs (e.g. fertilizers, credit), over which they may have little control, thereby increasing local food security.

Social and Institutional Dimensions

Land Tenure

Protecting soils and enhancing their fertility (land stewardship) implies ensuring productive capacity for future generations. However, in the quest to improve soil quality for the future (probably the single most important factor to determine whether farmers are interested in the issue) is whether they will benefit from the change. Security of land tenure is, therefore, an extremely important factor in this respect. If land security is not guaranteed, there is little reason for farmers to invest in a method that will bring them income in the future rather than immediate rewards.

Labour

If compared to large-scale mechanized agricultural systems, organic systems appear more labour-intensive. Using no external inputs requires that fertilizers and pest control methods be internally produced using labour intensive techniques. Labour requirements tend to be high on some organic farms (e.g. plantations), on organic farms where labour-intensive methods are used (such as strip farming, manual weeding, composting), and in low ecological potential areas that use techniques that require significant labour (such as Zai planting pits).

The timing of labour requirement throughout the year is an important aspect of labour. Care should also be paid to on farm multiple activities (e.g. agro-tourism, processing) and off-farm labour time as often part of the family income comes from non-farm and off-farm activities. Another important issue to consider is not the quantity of labour, but the quantity of output per unit of labour, or labour productivity. While organic agriculture is likely to generate good labour productivity, labour return depends on the available amount of labour at farm household level. Often, the increase of costs of labour to produce extra units of crops is higher than the increase in outputs.

In many areas, labour is a critical constraint in agricultural production, especially when considered on a seasonal basis. Achieving high levels of labour use may not be physically possible or economically viable on small-scale farms in developing countries and may also add to household drudgery, especially for women. This is also true for cash-poor farmers and those supplementing their incomes with off-farm work. Labour and total costs are generally lower on private organic farms. In the developed world, labour scarcity and costs may deter farmers from adopting organic management systems. However, where labour is not such a constraint, organic agriculture can provide employment opportunities in rural communities. Furthermore, the diversification of crops typically found on organic farms, with their various planting and harvesting schedules, may result in more work opportunities for women and a more evenly distributed labour demand which helps stabilize local/regional employment.

Justice

The concept of "fair trade" implies a concern of the buyer for social justice for those who work in agriculture, especially with regard to a "fair wage". In fair-trade projects, traders ensure that producers receive a minimum return for their produce irrespective of the actual market price. At present, certification that guarantees fair trade does not necessarily imply organic production. Organic certification organizations favourably consider inclusion of "reasonable wage conditions".

Improving the situation of women is recognized as an important issue within organic agriculture. The social environment of the workers engaged in organic agriculture, namely their working conditions and environment, often results in improved housing situations and childcare facilities. Availability of work, gender distribution of labour and access to knowledge are key considerations in organic agriculture.

Attitudes and Perceptions

The single biggest constraint to the development of organic agriculture is that most people in all kinds of areas (including scientists, researchers, extension officers and politicians) strongly believe that organic agriculture is not a feasible option to improve food security. For this reason, very few farmers can obtain information about this management system, even when they inquire about it. If those who make policy decisions on the allocation of resources, such as for research and extension, are not aware of the possibilities of organic agriculture, no positive consideration towards this farming system can be expected.

Investments and Benefits

The advancement of organic agriculture to date has to a large extent been due to private investment. This has been in the form of consumers' willingness to pay for organic commodities (price premiums) and farmers' readiness to experiment and innovate, despite the risks involved with such on-farm research. Many of the inputs used in organic agriculture are public goods (such as knowledge about practices) which can be used without impeding use by others. Hence, there is little private interest in promoting particular inputs, which are used in organic agriculture. Organic management, which relies on local knowledge of complex interactions and variations of conditions carries an enhanced potential for more equitable distribution and access to productive resources, namely land.

Knowledge

Empirical knowledge of natural processes takes a long time to consolidate itself. Traditional knowledge can be improved through selective introduction of results of modern science in areas such as energy flows and biogeochemical cycles, biotic and abiotic factors, which regulate plant development, renewable energy technologies, and management techniques. The integration of traditional technologies with emerging technologies would lead to an improvement of traditional technologies in terms of: i) cost per unit; ii) productivity per unit of factor input; and iii) quantity and quality of output.

Research

The conservation of ecological foundations (soils, aquifers, forests, biodiversity, and the climate systems) is essential for continuous advances in crop and animal productivity. Conventional research on management of natural resources for agriculture (i.e. soil, water, biological resources, and related nutrient and energy balance) is not responding to farmers' needs. To be successful, agricultural innovations should be based on an interactive process between farming communities and agricultural research institutions. Applied and adaptive agricultural research could be successfully implemented through location-specific and on-farm research, jointly with rural and tribal families, where partnerships, extension, and communication are integrative parts of the process of research and knowledge sharing.

Community Participation

Engaging in organic production means experimenting new techniques, introducing different management of labour time, investing efforts in different management of space, adapting and refining solutions to change, comparing different options with farmers that have similar conditions, and making appropriate choices. This can only be achieved through farmers' participation in research and its application. This on-farm research component can support rural communities, and generate new knowledge that will benefit all farmers. Consistent labour needs, combined with the enhanced capacity of the land and protection of water associated with organic agriculture, may encourage people to permanently locate and thus reinvigorate rural communities. The establishment of cooperation between farmers is instrumental in helping farmers to become a stronger and more independent partner in the agro-business environment. In addition, providing a critical mass for renewed rural community structures sets an end to the isolation of farmers, thus increasing the viability of rural life. Most importantly, various forms of cooperation within the food chain are necessary to overcome the gap between farmers and consumers.

Self-reliance

Within organic agriculture, the use of locally available inputs is encouraged. The effect on the local community of such a form of agriculture is, therefore, likely to be greater than when inputs are imported from outside the community. In those cases where synthetic fertilizers and pesticides are imported, adoption of organic agriculture techniques means a decrease in imports, decreasing the need for foreign currency and credit. The site-specific nature of organic agriculture also means that indigenous species and knowledge, so often discounted, are of great value. In many places, this knowledge has been eroded with the introduction of high external input agriculture, promotion of monoculture, and selection of "improved products." Farmers may readily welcome a management system close to their own traditions and not driven solely by a production ethic.

Policies and Regulations

Norms, standards, inspection and certification in organic agriculture have been established and remain enacted by the private sector. Recently, a number of countries have established guidelines, regulations and policies to assist producers to respond to the growing demand for organic food and to control fraudulent practices. Favourable policies and institutional structures are important incentives to the adoption of organic agriculture. Where supportive government policies are lacking, and especially where inspection organizations are unavailable (or need to be imported at high expense), interested communities need to mobilize and organize themselves to procure necessary inputs and to access markets.

References

- Organic-farming-beginners-guide: kisancentral.com, Retrieved 1 May, 2019
- Transition, organic, orgfarm-successful, org-farm: tnau.ac.in, Retrieved 4 Feb, 2019

- Principles-organic-agriculture, organic-landmarks: ifoam.bio, Retrieved 4 January, 2019

- Organic-farming, health: ndtv.com, Retrieved 13 Febuary, 2019

- An-introduction-to-cover-crop-species-for-organic-farming-systems:extension.org, Retrieved 17 March, 2019

- Organic-farming-in-india: techgape.com, Retrieved 1 May, 2019

- Advantages-and-disadvantages-of-organic-farming-in-modern-agriculture: greengarageblog.org, Retrieved 25 August, 2019

- Organic-food, topic: britannica.com, Retrieved 31 July, 2019

- Organic-foods, healthy-eating: helpguide.org, Retrieved 17 February, 2019

- Organic-certification-of-vegetable-operations: extension.org, Retrieved 31 July, 2019

Chapter 2

Types of Organic Farming

There are a number of types of organic farming such as organic aquaculture, organic horticulture and organic gardening. Vegan-organic farming is a type of organic farming which seeks to cause minimal harm to animals. This chapter discusses in detail the techniques and methods used in these types of organic farming.

Organic Aquaculture

Organic Aquaculture is the only solution to increase fish production in sustainable and environment friendly manner.

"Organic aquaculture is production of high quality foods in a stable aquatic ecosystem by managing the natural resources and environment without any negative effects and to secure the genetic diversity and richness of species in a native system."

Current problem with the industrial aquaculture practice of fish harvested from wild as feed for the production of cultured fish, 3 tons of wild fish is used to produce feed for the production of 1 ton of farmed fish, so this depletes the natural stock available in wild. To increase production, fast growing exotic fish varieties are farmed this result in weakening of the native species and transfer of disease from farmed aquatic animals to wild fish is also major problem in the current aquaculture systems. Organic aquaculture is a method to reduce the abovementioned adverse effects of the industrial aquaculture practice. Organic aquaculture is most important in the sustainable and environmental friendly aquaculture production. This method of culture also farms the aquatic organisms in conditions similar to that of the natural environment.

As in case with the other forms of food production industries there is some consumer interest in organic aquaculture. However fish farmers have been slow to adopt the organic standard as many claim that modern aquaculture practices are already " organic " in principal but do not meet the strict legal standards.

Principles of Organic Aquaculture

The main principles of organic aquaculture are as follows:

- Monitoring of environmental impact.
- Natural breeding procedures without use of hormones and antibiotics.
- No use of inorganic fertilizers.

- Integration of natural plant communities in farm management.

- No synthetic pesticides and herbicides.

- Feed and fertilizer from certified organic agriculture and fisheries.

- Organic criteria of sustainability for fishmeal sources.

- Absence of GMOs(Genetically Modified Organisms) in stocks and feed.

- Stocking density limits.

- Restriction of energy consumption (e.g. regarding oxygenation).

- Preference for natural medicines.

- Processing in approved organic facilities.

Organic Horticulture

Organic horticulture is the science and art of growing fruits, vegetables, flowers, or ornamental plants by following the essential principles of organic agriculture in soil building and conservation, pest management, and heirloom variety preservation.

Vermicompost, mulches, cover crops, compost, manures, and mineral supplements are soil-building mainstays that distinguish this type of farming from its commercial counterpart. Through attention to good healthy soil condition, it is expected that insect, fungal, or other problems that sometimes plague plants can be minimized. However, pheromone traps, insecticidal soap sprays, and other pest-control methods available to organic farmers are also sometimes utilized by organic horticulturists.

Horticulture involves five areas of study. These areas are floriculture (includes production and marketing of floral crops), landscape horticulture (includes production, marketing and maintenance of landscape plants), olericulture (includes production and marketing of vegetables), pomology (includes production and marketing of fruits), and postharvest physiology (involves maintaining quality and preventing spoilage of horticultural crops). All of these can be, and sometimes are, pursued according to the principles of organic cultivation.

Organic horticulture (or organic gardening) is based on knowledge and techniques gathered over thousands of years. In general terms, organic horticulture involves natural processes, often taking place over extended periods of time, and a sustainable, holistic approach - while chemical-based horticulture focuses on immediate, isolated effects and reductionist strategies.

Organic practices arise from the understanding that all organisms in nature are interdependent, and in order to have healthy plants we must foster the health of their entire ecosystem. These practices go beyond integrated pest management, beyond the use of so-called organic fertilizers and pesticides. They acknowledge the concept of intrinsic health, and seek to create environments that cater to the well-being of all their inhabitants.

Organic Gardening Systems

There are a number of formal organic gardening and farming systems that prescribe specific techniques. They tend to be more specific than, and fit within, general organic standards. Biodynamic farming is an approach based on the esoteric teachings of Rudolf Steiner. The Japanese farmer and writer Masanobu Fukuoka invented a no-till system for small-scale grain production that he called Natural Farming. French intensive and biointensive methods and SPIN Farming (Small Plot Intensive) are all small scale gardening techniques.

A garden is more than just a means of providing food, it is a model of what is possible in a community - everyone could have a garden of some kind (container, growing box, raised bed) and produce healthy, nutritious organic food, a farmers market, a place to pass on gardening experience, and a sharing of bounty, promoting a more sustainable way of living that would encourage their local economy. A simple 4'×8' (32 square feet) raised bed garden based on the principles of bio-intensive planting and square foot gardening uses fewer nutrients and less water, and could keep a family, or community, supplied with an abundance of healthy, nutritious organic greens, while promoting a more sustainable way of living.

Other methods can also be used to supplement an existing garden. Methods such as composting, or vermicomposting. These practices are ways of recycling organic matter into some of the best organic fertilizers and soil conditioner. Vermicompost is especially easy. The byproduct is also an excellent source of nutrients for an organic garden.

Soil

"Soil is the result of interaction between three equal partners: the a-biotic components of the soil, living organisms and environmental components (temperature, water, air). Each of these brings different assets to the partnership, and each makes a unique value contribution to the whole. The synthesis of all contributions - soil - is greater than the sum of its parts. Soil must be understood and managed from this holistic perspective."

"The primary a-biotic component of soil is rock which, as it becomes reduced to smaller particles, contributes the secondary components: sand, silt, clay and mineral nutrients." The "combination of particle size, shape and arrangement affects the soil's compactability, air supply and water holding capacity."

Living organisms make the following contributions to the soil partnership: conversion of carbon and nitrogen gas into solid compounds, modification of inherent soil fertility through biological transmutation, conversion of mineral soil nutrients into organic forms, creation of multi-level food production and storage systems, and creation of habitat through structural improvement of the soil.

Environmental components of soil include air, water and temperature. "The environmental components literally become incorporated into living organisms, and into the soil ecosystem. They are not just 'factors' or 'influences', they contribute essential building blocks and energy to fuel the processes. They are vital and active participants."

Soil is an ecosystem "created through the proportionate availability and interaction of all three soil partners. Soil is the habitat living organisms have synergistically created for themselves. Our job is to support them, not to put them out of work".

Managing the soil is very important. If your garden is healthy then insects will not attack the plants. Insects only attack plants that are unhealthy. To keep your garden healthy give it organic matter and humus to survive. The most important thing is to give your garden lots of attention and your energy.

Nutrient Management

Principally, animal manures, compost, green manures, bio-fertilizers, mixed organic fertilizers are used in organic farming. Nitrogen is provided by legume crops having nitrogen fixing symbiotic bacteria and by soil habiting non-symbiotic bacteria. Enrichment of phosphorus in soil is done by incorporation of rock-phosphate, VAM (Vesicular Arbuscular Mycorrhiza - soublises phosphorus for greater availability to the plants) and VAM treated compost. Potassium is provided by wood ash, sea weeds, tobacco stem; used alone or in combination with others. Beside it permanent mulching layer reduces the potassium leaching.

Apply lime 2-3 months before planting to correct soil acidity.

Pest Control Approaches

Differing approaches to pest control are equally notable. In chemical horticulture, a specific insecticide may be applied to quickly kill off a particular insect pest. Chemical controls can dramatically reduce pest populations in the short term, yet by unavoidably killing (or starving) natural control insects and animals, cause an increase in the pest population in the long term, thereby creating an ever increasing problem. Repeated use of insecticides and herbicides also encourages rapid natural selection of resistant insects, plants and other organisms, necessitating increased use, or requiring new, more powerful controls.

In contrast, organic horticulture tends to tolerate some pest populations while taking the long view. Organic pest control requires a thorough understanding of pest life cycles and interactions, and involves the cumulative effect of many techniques, including:

- Allowing for an acceptable level of pest damage.

- Encouraging predatory beneficial insects to flourish and eat pests.

- Encouraging beneficial microorganisms.

- Careful plant selection, choosing disease-resistant varieties.

- Planting companion crops that discourage or divert pests.

- Using row covers to protect crop plants during pest migration periods.

- Rotating crops to different locations from year to year to interrupt pest reproduction cycles.

- Using insect traps to monitor and control insect populations.

Each of these techniques also provides other benefits, such as soil protection and improvement, fertilization, pollination, water conservation and season extension. These benefits are both complementary and cumulative in overall effect on site health. Organic pest control and biological pest control can be used as part of integrated pest management (IPM). However, IPM can include the use of chemical pesticides that are not part of organic or biological techniques.

Pest, Disease and Weed Management

Suitable crop rotations, green manuring, use of balanced fertilizers, proper care during nursery, mulching etc. play important role to protect the crop from insect pest and disease. The following practices are recommanded:

1. Use trichoderma for seed treatment.

2. Use resistant varieties.

3. Use disease free planting material.

4. Use mulching for those vegetables whose fruits touches the soil.

5. Clean the crop residue.

6. Summer fallowing and flooding.

7. Use spray Bt, NPV, Beauveria.

8. Use insect traps- such as pheromone trap, light trap, yellow trap, sticky trap.

9. Use botanicals like neem, garlic etc.

10. Use trap crops to misguide the insect and protect the main crop.

Organic Gardening

Organic gardening is essentially gardening without using synthetic products like fertilizers and pesticides. It involves the use of only natural products to grow plants in your garden. Organic gardening replenishes the natural resources as it uses them. In organic gardening, you consider your plants as part of the larger natural system that begins with the soil and includes water supply, the wildlife; insects and people. Everyone wants the food we serve to our families as well as our environment to be safe and healthy. A good organic gardener strives to ensure that his or her activities are in harmony with the natural ecosystem and aims at minimizing exploitation as well as replenishing all the resources consumed by his or her garden.

Gardeners and people who have come across the word organic gardening probably usually desire to know what it means. Organic gardening is a terminology that simply refers to growing of plants, vegetables, and fruits in the best natural way without the use of pesticides or synthetic chemical fertilizers. Even so, organic gardening is more than simply avoiding the use pesticides and synthetic fertilizers.

Gardening organically encompasses supporting the health of the entire gardening system naturally. It means working in harmony with the natural systems including the soil, water supply, people, and even insects with an ultimate aim of minimizing destruction to living and non-living things in the natural environment while constantly replenishing any resources utilized during gardening.

Organic gardening fundamentals requires cultivation emphasized on creating an ecosystem that nourishes and sustains soil microbes and plants, while also benefiting insects rather than just putting seeds in the ground and letting them grow.

To begin with, there are three major areas to concentrate on to maintain the objectives of organic gardening. These includes: soil management which is dealt with by using organic fertilizer; weed management which is managed by manual labor and use of organic ground coverings; and lastly pest control which is dealt with by promoting beneficial insects and companion planting. These are the top key strategies for becoming an organic gardener. Proper knowledge is essential in organic gardening and requires simple fundamental lessons to get reliable results.

Small-organic-garden

To sum up, organic gardening is as simple as relying on intermingling plant types and varieties, use of companion planting, dense planting in order for some plants to offer companion to vulnerable plants, and supporting natural systems to minimize spread of pests and diseases.

Process to Start an Organic Garden

The following is a step by step process on how to start an organic garden:

Prepare the Soil

The soil is the most important thing or resource when it comes to organic gardening. It is achieved through continuous addition of organic matter to the soil by using locally available resources in every possible aspect. If you want your plants to be healthy, you will need to thoroughly prepare the soil on which they will grow on.

Just like human beings, plants require food and the food in this case comes from the soil. Therefore, you need to ensure that your plants get plenty of fresh nutrients. Proper soil conditioning will give your plants all the nutrients they need.

Chemical soil treatments not only destroy the soil composition but they also harm the important microorganisms, worms, and bacteria in the soil. To begin, you will have to test the soil PH. You can do this by buying a home testing kit or simply collect some soil samples and send them to the local agricultural extension office for proper testing and analysis.

Make Good Compost

As you wait for soil sample results, you can make the compost. Compost helps in providing plants with nutrients, helps conserve water, helps in the reduction of weeds and helps in keeping food as well as yard waste out of landfills.

Compost can be obtained or made from locally available resources such as leaves, grass trimmings, yard garbage/remains, and kitchen waste. Alternatively, compost is readily available for purchase from mulch suppliers or organic garden centers. You can use these steps in making compost.

- Measure out space that is at least three square feet.

- Get a pile of natural dead plants or leaves.

- Add alternating layers of leaves, garden trimmings (carbon) and nitrogen (green) materials, for instance, kitchen leftovers and manure. Put a layer or separate them with a layer of soil.

- Cover the pile with about 4-6 inches of soil.

- Turn the pile every time a new layer is added to the mixture. During this process, ensure that water is added to keep the mixture moist to enhance microorganism activity.

- The compost should not smell but in case it does, add some more dry leaves, saw dust or straw and then turn it regularly. Do this for about four weeks or a month and you will have good compost needed for your organic garden.

Prepare your Garden

Once you are waiting for the compost to be ready, the next step is to prepare your garden. After receiving a go ahead from your local agricultural extension officer concerning the right soil type, it is now time to prepare the garden area. Using available gardening tools you can carefully prepare your garden. However, it is important to ensure that you do not completely destroy the soil.

Choose the Right Plants

Once you are through with preparing your garden, the next step is to select the right plants for your garden. Soil sampling and testing will come in handy at this stage. It is important to choose plants that will thrive well in specific micro-conditions of your soil type. Carefully choose plants that will thrive well in different spots in your garden in terms of moisture, light, and drainage as well as the soil quality. Remember the healthier your plants are, the more and more resistant your crops will be to attackers.

Another mechanism for growing organically is to select plants suited to the garden. Crops well adapted to the gardens climate and conditions are better able to grow with minimal input. Also, growing crops well adapted to the site ensure greater natural defenses. Meaning, little attention and input is required for boosting crop productivity.

When purchasing seedlings, ensure that you go for plants that are raised without synthetic chemicals or pesticides. Your local farmers market is a good place to do your purchases. You will not only find a variety of plants but plant varieties that are best suited for your local area. Carefully select plants that look healthy and are without overcrowded roots.

Plant the Crops in Beds

When planting your crops ensure that you plant them in wide beds. Planting them in beds prevents you from walking on them and destroying the soil surface when harvesting or when cutting the flowers. Additionally, grouping crops helps in reducing weeding, wastage of water and makes it easier for you to apply compost. It also enables the plants to utilize the available nutrients and water. Ensure that there is adequate space between the rows. This helps in promoting air circulation which helps in repelling fungal attacks.

Water the Crops

Once you have planted your crops, the next step is to water them. It is good to water the plants immediately after planting them to give them the much-needed water to enable them to continue growing. You can also water them every morning. It is recommended to water your pants in the morning because there are no strong winds, mornings are cool and the water lost as a result of evaporation is immensely reduced. Experts recommend considerable, infrequent watering for plants that are already established.

Weeding

According to CSU, Weeds reduce crop yield by competing for water, light, soil nutrients, and space. In agricultural crops, weeds can reduce crop quality by contaminating the commodity. They can serve as hosts for diseases or provide shelter for insects to overwinter. It may be a hard work to pull out weeds by hand but then its a good exercise that helps you to get some fresh air.

Provide Nutrients to your Plants

When you do organic gardening, you need to look out for eco-friendly ways to protect your plants from toxic pesticides and fertilizers. So, you need to make sure that your plants get enough light, nutrients and moisture that help them to grow better. Also, a diverse garden would help to prevent pests, by limiting the amount of one type of plant and gives boost to biodiversity.

Instead of using chemical pesticides, organic gardening emphasizes on promoting beneficial insects and companion planting. Organic farmers need not to eliminate insects and diseases by using chemicals. Rather, pest control is done through keeping pests and diseases below damaging levels. One of the major mechanisms is by promoting beneficial insects and pest predators such as bats, birds, lizards, toads, and spiders.

The key to succeeding in this area is to grow a wide variety of companion crops that supports the ecological niche of these species. Avoiding the use of synthetic pesticides as well ensures their survival. Nipping off infected/infested leaves or buds, uprooting infected crops, crop rotation, and handpicking insect pests and eggs are excellent methods for controlling pest populations. Maintaining garden cleanliness is another effective tool for organic pest control.

Natural sprays or pesticides can equally be used in addition to the cultural pest control methods. They are readily available from organic garden centers and their products contain the bacterium Bacillus, neem oil, and minerals like copper. Their ability to break down quickly has promoted their wide usage in place of the synthetic chemical pesticides. Besides, there are some vegetable/ fruit pests and diseases that are beyond natural and organic control, thereby requiring the use of natural sprays.

Use of Organic Fertilizers

As much as organic matter and compost will improve water and nutrient retention in the soil, the supply of all the required nutrients for healthy and productive growth is limited. Besides compost, organic gardening requires additional fertilizers drawn from natural sources such as: plant products like wood ash; natural deposits for example rock phosphate; and animal by-products as well as manures.

Agricultural lime is another natural product frequently added to soil to improve its quality. It is produced from naturally occurring limestone and is added to the soil to optimize pH if the soil is too acidic. Soil pH levels vary from one locality to another. Local extension offices usually provide guidance on soil pH level testing within their jurisdiction areas. However, most soils do not need additional liming.

Care of Organic Garden

There are important maintenance tips for your organic garden if you want to maintain or harvest healthy plants. Some of these practices include:

- Mulching.

- Watering in the mornings.

- Use of compost.

- Use of natural manure especially from animals that do not eat meat.

- Weeding at the right time.

- Prune regularly to allow for proper aeration and effective use of nutrients including light.

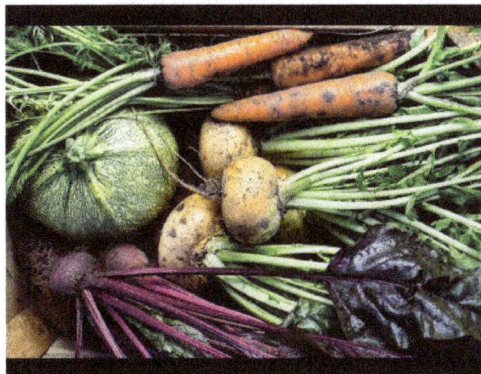

Various Benefits of Organic Gardening

Reduces the Amount of Pesticides you and your Family Consume

Organic gardening focuses on the use of only natural products to grow plants. This means that there is no use of pesticides. Therefore, the crops obtained from this type of gardening are free of pesticides and other chemicals. You will live a healthy life without worrying about consuming chemicals through consuming crops.

Organic gardening promises good health because the produce is free from toxic ingredients and other synthetically enhanced chemicals. The fruits and vegetables grown in the organic garden do not have the chemical residues which enter the body when eaten. Organic veggies are also proven to contain higher mineral and vitamin contents compared to those grown using pesticides, herbicides, and chemical fertilizers.

Furthermore, complementary medicine professionals affirm that there are high nutrient concentrations such as Vitamins C and D in organic food products. Organic gardening also adds an extra benefit of exercising the body, especially from manual labor including planting, weeding, and harvesting. Thus, gardening can be an enjoyable and meaningful way to boost physical exercise. The outdoor environment also offers a refreshing means of connecting with nature, sunshine and fresh air, acting as a stress reliever

Helps in Conserving the Environment

Use of chemicals on crops is one way in which we pollute the environment. The chemicals sprayed or applied on the crops seethe through the soil into the water. This puts the microorganisms at risk. When spraying the crops, the wind carries away the chemicals into the atmosphere, and this is air pollution. Therefore, embracing organic gardening is one of the best ways of maintaining a healthy environment.

Choosing organic gardening massively aids in environmental preservation. Growing vegetables and fruits in a natural way will not only ensure healthy produce but also promote a friendly and toxicity free environment. Organic gardening is the surest way of preserving a healthy and green environment by ensuring ecological balance and minimal disturbance to the natural environment.

It ensures birds, small animals, and beneficial insects are free from chemical harm. Organic matter used to prepare the soil also helps to improve the soil quality. Therefore, organic gardening offers the most beneficial outcome for good environmental health. Putting an end to chemical fertilizers and pesticides that leach into the ground and find way into water supply is only a possibility through organic gardening.

Reduces Greenhouse Gas Emissions

Crops that contain chemicals release these chemicals into the atmosphere in the form of greenhouse gasses. These gasses combined with other impurities in the air pose a risk to the air we breathe in. Additionally, these gasses also contribute immensely to global warming, and we all know the negative effects of global warming.

Better Taste

Organic vegetables and fruits contain scent and taste which are simply better due to their natural growth compared to those grown commercially. To a very large extent, veggies and fruits grown commercially cannot bear or beat the natural flavors of those that are grown organically. Fresh vegetables and fruits from the garden have always tasted better and have natural flavor.

Saves Money

A tremendous way of saving money is by growing your own organic vegetable garden. Saving money is something that every person wants to do but it can only be realized by undertaking small initiatives like growing own veggies and fruits. Through organic gardening, one can save up to 50% of the money used to buy fruits and veggies at supermarkets as well as other perishable stores.

To conclude, organic gardening is beneficial to the people and the environment. However, learning how to start an organic is a very important step in ensuring that you conserve the environment and that you grow your own delicious fresh produce. To give your crops the needed nourishment, you can use natural fertilizers, for example, seaweed extracts, fish emulsion or manures obtained from animal droppings especially cow and chicken droppings that is readily available or bought from your local garden center.

Vegan-organic Gardening

Vegan organic gardening and farming is the organic cultivation and production of food crops and other crops with a minimal amount of exploitation or harm to any animal.

Vegan-organic gardening avoids not only the use of toxic sprays and chemicals, but also manures and animal remains. Just as vegans avoid animal products in the rest of our lives, we also avoid using animal products in the garden, as fertilizers such as blood and bone meal, slaughterhouse sludge, fish emulsion, and manures are sourced from industries that exploit and enslave sentient beings. As these products may carry dangerous diseases that breed in intensive animal production operations, vegan-organic gardening is also a safer, healthier way to grow our food.

In veganic growing situations, soil fertility is maintained using vegetable compost, green manures, crop rotation, mulching, and other sustainable, ecological methods. Occasional use of lime, gypsum, rock phosphorus, dolomite, rock dusts and rock potash can be helpful, but we try not to depend on these fertilizers as they are non-renewable resources.

Soil conditioners and fertilizers that are vegan-organic and ecologically sustainable include hay mulch, wood ash, composted organic matter (fruit/vegetable peels, leaves and grass clippings), green manures/nitrogen-fixing cover crops (fava beans/clover/alfalfa/lupines), liquid feeds (such as comfrey or nettles), and seaweed (fresh, liquid or meal) for trace elements.

A border of marigolds helps to deter certain insects, and they also have a root system that improves the soil.

Composted Organic Matter

A compost pile consists of food waste such as fruit and vegetable rinds, that is covered by course material like leaves or grass clippings. The object is to create layers of food material alternating with covering material to allow aeration. When a bin is full, the pile is flipped and covered by black plastic or weed mat to protect it from rain and create heat. It can be flipped again after a period of time, so the bottom becomes the top. Cover again and within a couple of months, depending on the climate, nature's master recycling plan will have taken its course and you will have vitamin-rich soil.

Green Manures and Nitrogen-fixing Crops

Green Manure is a cover crop of plants, which is grown with the specific purpose of being tilled into the soil. Fast-growing plants such as wheat, oats, rye, vetch, or clover, can be grown as cover crops between gardening seasons then tilled into the garden as it is prepared for the next planting. Green manure crops absorb and use nutrients from the soil that might otherwise be lost through leaching, then return these nutrients to the soil when they are tilled under. The root system of cover crops improves soil structure and helps prevent erosion. Nitrogen-fixing crops such as vetch, peas, broad beans (fava beans) and crimson clover add nitrogen to the soil as they are turned under and decompose. Cover crops also help reduce weed growth during the fall and winter months.

Liquid Feeds such as Comfrey or Nettles

Fill a container with grass cuttings, nettles, weed or comfrey leaves. Cover with water at a rate of one part brew to three parts water. Cover the container, and leave for two to four weeks. Preferably strain out (through an old stocking) the weed seeds and plant material that will block up the spout of your watering-can. Nettles give the best multi-purpose feed and comfrey alone will give a feed rich in potash.

Hay Mulches

Using a thick layer of hay to cover the earth feeds the soil with organic matter as it breaks down. It also suppresses weeds and encourages worms to live in your soil. When putting gardens to sleep over the winter, cover them with a very thick layer of hay mulch.

Seaweed Fresh, Liquid or Meal

Used for trace elements. Seaweed is best harvested fresh from the sea as opposed to washed up and sitting on beaches. Some veganic gardeners use bulk spirulina or kelp meal (used for potash and trace minerals).

Worm Castings Vermiculture, Vermicastings and Vermicomposting

Re-establish natural worm populations in your garden. Composting worms love cool, damp and dark environments (like under black weed mat or a thick layer of hay mulch), and will breed optimally when these conditions are maintained. Worm castings are a rich, all-natural source of organic matter with lots of nutrients and moisture-holding capabilities. Earthworm castings are known to have an extraordinary effect on plant life. They improve the soil structure and increase fertility.

Lime

The primary purpose for using lime in the garden is to reduce the acidity of the soil, otherwise known as raising the pH level or 'sweetening the soil'. Most plants prefer a fairly neutral soil for optimum growth. You can have your soil tested to see if it is acidic or alkaline. Lime also enriches the soil with calcium and magnesium. Calcium is essential for strong plant growth and aids in the absorption of other nutrients. Lime can also be used for breaking up heavy clay soil.

Gypsum Hydrated Calcium Sulfate

Gypsum is also used where more calcium is needed, but unlike Lime, it enriches the soil without raising the pH level.

Neem

Known as the wonder tree in India, Neem has been in use for centuries in Indian agriculture as the best natural pest repellent and organic fertilizer with insect sterilization properties.

EM Bokashi

Bokashi is a Japanese term that means 'fermented organic matter'. EM means Effective Micro-organisms and consists of mixed cultures of naturally occurring, beneficial micro-organisms such as lactic acid bacteria, yeast, photosynthetic bacteria and actinomycetes. It is a bran-based material that has been fermented with EM liquid concentrate and dried for storage. Add to compost to aid in the fermentation of the organic matter. EM Bokashi should be stored in a warm, dry place out of direct sunlight.

Green Sand

A soil amendment and fertilizer: It is mined from deposits of minerals that were originally part of the ocean floor. It is a natural source of potash, as well as iron, magnesium, silica and as many as 30 other trace minerals. It may also be used to loosen heavy clay soils. It has the consistency of sand but has 10 times the ability to absorb moisture.

Alfalfa Meal, Flax Seed Meal, Cottonseed Meal and Soya Meal

Sources of nitrogen.

Epsom Salts: An excellent source of magnesium.

Dolomite: A finely ground rock dust which is the preferred source of calcium and magnesium.

Rock Phosphate: Phosphorus is an essential element for plant and animal nutrition. It is mined in the form of phosphate rock, which formed in oceans in the form of calcium phosphate called phosphorite. The primary mineral in phosphate rock is apatite.

Rock Dusts Stonemeal

Used to re-mineralize soil that has become depleted through industrial and agricultural practices. It releases slowly into the soil and can be applied directly, in combination with other fertilizers, or added to the compost. These products have a highly stimulating effect on microbial activity.

Rock Potash Potassium or Wood Ash

Potassium is an essential nutrient that enhances flower and fruit production and helps 'harden' foliage to make it less susceptible to disease. Rock potash is very slow-acting. It releases gradually as it weathers, which can take years. Use it when preparing soil before planting.

References

- View, mod: iasri.res.in, Retrieved 25 August, 2019
- Organic-farming-horticulture, content: agropedia.iitk.ac.in, Retrieved 31 July, 2019
- Start-an-organic-garden: conserve-energy-future.com, Retrieved 25 August, 2019
- Beginners-guide-to-veganic-gardening: gentleworld.org, Retrieved 20 July, 2019

Chapter 3

Weed and Pest Management in Organic Farming

Pest management in organic farming is done by manipulating agroecosystem processes. Weed control is an important part of pest control that aims to stop weeds from growing in desired flora and fauna. This chapter discusses in detail the theories and methodologies related to organic weed and pest management.

Insect Pest Management in Organic Farming System

Organic agriculture is a holistic production system that sustains the health of soils, ecosystems, and people. It relies on ecological processes, biodiversity, and cycles adapted to local conditions rather than the use of inputs with adverse effects. Organic agriculture combines tradition, innovation, and science to benefit the shared environment and promote fair relationships and a good quality of life for all involved. Holistic means near-closed nutrient and energy cycle system considering the whole farm as one organism. Organic agriculture relies on a number of farming practices based on ecological cycles and aims at minimizing the environmental impact of the food industry, preserving the long-term sustainability of soil and reducing to a minimum use of non-renewable resources. Organic agriculture is both a philosophy and a system of farming aiming to produce food that is nutritious and uncontaminated with substances that could harm human health. Organic farming benefits to the ecosystem include conservation of soil fertility, carbon dioxide storage, fossil fuel reduction, preserving landscape, and preservation of biodiversity.

Pest management in organic farming is achieved by using appropriate cropping techniques, biological control, and natural pesticides (mainly extracted from plant or animal origins). Weed control, the main problem for organic growers, can be managed through cultural practices including mechanic cultivation, mulching, and flaming. Organic farming is characterized by higher diversity of arthropod fauna and conservation of natural enemies than conventional agriculture.

According to the IFOAM, organic agriculture is guided by four principles: health (soil, plant, animal, and human), ecology (living ecological systems and cycles), fairness (environment and life opportunities), and care (protect the health and well-being of current and future generations as well as the environment). The US Congress passed the organic food product act in 1990, while the European Union (EU) set up the first regulations on organic farming in 1991, and in the same year, the Codex Alimentarius Commission officially recognized organic agriculture. Gomiero et al. gave more details on history of organic farming, total global areas, organic standards, and impact on the environment.

Principles and Strategies of Crop Protection in Organic Farming System

Pest management in organic farming is a holistic (whole-farm) approach that largely depends on the ecological processes and biodiversity in the agroecosystem. Accordingly, most IPM tactics, principles, and components match with organic farming systems. The goal of this strategy is to prevent pests from reaching economically damaging levels without causing risk to the environment. Successful IPM programs in organic farming may have the following components: (1) monitoring crops for pests, (2) accurately identifying pests, (3) developing economic thresholds, (4) implementing integrated pest control tactics, and (5) record keeping and evaluation.

The factors that render crop habitat unsuitable for pests and diseases include limitation of resources, competition, parasitism, and predation. These factors play an important role in maintaining equilibrium of the agroecosystem and suppression of harmful pests. Faunal and floral diversities play a substantial role in pest and disease management in organic farming system.

Differences between Organic and Conventional Farming with Respect to Plant Protection

Few options of plant protection substances are available for certified organic growers compared to conventional ones. Thus, they should capitalize on the natural processes and management of the ecosystem to control harmful organisms. Organic farms had a more diverse arthropod fauna, on average, than conventional farms. The average for five 30-second vacuum samples per farm was approximately 40 arthropod species in conventional tomato compared to 66 species in organic tomato fields. Additionally, natural enemies (parasitoids plus predators) were more abundant on organic farms. Arthropod biodiversity, as measured by species richness, was, on average, one-third greater on organic farms than on conventional farms.

Under organic farming systems, the fundamental components and natural processes of ecosystems, such as soil organism activities, nutrient cycling, and species distribution and competition, are used directly and indirectly as farm management tools to prevent pest populations from reaching economically damaging levels. Soil fertility and crop nutrients are managed through tillage and cultivation practices, crop rotations, and cover crops and supplemented with manure, composts, crop waste material, and other allowed substances.

Soil-borne and root pathogens are usually found in low levels in organic farming as compared to conventional farming. Pathogens such as Pythium spp., Sclerotium rolfsii, Phytophthora spp., and some Fusarium can survive on organic matter of the soil, in the absence of their hosts for long periods, and are thus difficult to be controlled with crop rotation. Additionally, airborne pathogens cannot be controlled with cultural practice such as crop rotation. Powdery mildew and rust diseases (airborne) and insect pests such as aphids and whiteflies (sucking insects) are less serious in organic farming than in conventional farming due to lower nitrogen concentrations in foliar tissues or phloem of plants in the former compared with the latter. Almost all pesticides available for organic farming have short residual effects and work through direct contact mode of action as compared to the persistent systemic pesticides used in conventional farming. Table gives the main differences between organic and conventional farming with respect to soil fertility, biodiversity, and other criteria.

Table: Fundamental differences between organic and conventional farming.

Organic farming (OF)	Conventional farming (CF)
Synthetic fertilizers and synthetic pesticides are not permitted.	Synthetic fertilizers and synthetic pesticides are allowed.
Genetically modified organisms (GMOs) are not allowed.	GMOs can be used.
Soils have higher water holding capacity than CF.	Soils have less water holding capacity than OF.
OF has larger floral and faunal biodiversity than CF (complex crop pattern).	CF has smaller biodiversity than OF (simple crop pattern).
The agricultural landscape is characterized by heterogeneity (multicultural system).	The agricultural landscape is characterized by homogeneity (monocultural system).
Minimizing the use of nonrenewable resources by recycling plant and animal waste into the soils (on-farm inputs).	Depends largely on nonrenewable resources (off-farm inputs).
OF is more sustainable than CF.	CF is less sustainable compared to OF.
Strictly regulated by international and national institutional bodies such as Codex Alimentarius and IFOAM.	Not strictly regulated.
Crop protection depends mainly on natural processes such as soil fertility, crop cycle, and biodiversity (more preventive).	Crop protection relies mainly on human intervention with synthetic chemicals (more curative)

Crop Protection Practices in Organic Farming

Practices and tactics used in organic farming are based on the three management strategies, which include prevention, monitoring, and suppression. These practices will be intensively discussed in the following paragraphs:

Identification and Monitoring of Crop Pests

Crop pests include insects, weed, plant pathogens, invertebrate, and vertebrate animals. Identification of insect pests and their natural enemies is an important step in any pest management program. Insect pests and natural enemies could be identified using keys and field guides or otherwise consulting an official identification bodies. Unlike insect pests, plant pathogens including fungi, bacteria, virus, and nematodes are difficult to identify in the field and may need laboratory diagnosis. However, signs of insect damage and symptoms of plant diseases may be easily distinguished in the field. Weeds could be easily identified using key and field guides.

Monitoring is the regular inspection or scouting of field crops for pests, including insects, pathogens, nematodes, and weeds, to determine their abundance and level of damage. It serves as an early warning system for the presence of pests and diseases providing information for decision-making regarding management action and evaluation of control methods. Insect pests can be monitored through visual observation, pheromone and light traps, sticky traps, water traps, yellow traps, sweep nets, beating trays, and pitfall traps. Scouting data are used to develop economic thresholds, a useful decision-making tool to start control action when a pest population reaches or exceeds the specified economic threshold.

Tactics used for Pest Prevention and Suppression in Organic Farming

A successful integrated pest management (IPM) program in organic farming incorporates a variety of pest management tactics such as cultural, mechanical/physical, biological, and bio-pesticide (allowed for organic use) tactics individually or in combination. Each control tactic, discussed below, employs a different set of mechanisms for preventing and suppressing pest populations.

Cultural Pest Control

The goal of cultural control is to alter the environment, the condition of the host, or the behavior of the pest to prevent or suppress an infestation. It disrupts the normal relationship between the pest and the host and makes the pest less likely to survive, grow, or reproduce. In agricultural crops, crop rotation, selection of crop plant varieties, timing of planting and harvesting, irrigation management, crop rotation, and use of trap crops help reduce populations of weeds, microorganisms, insects, mites, and other pests. These cultural practices are more preventive than curative and thus may require planning in advance. The diversified habitat provides these parasites and predators with alternative food sources, shelter, and breeding sites. Tillage can cause destruction of the insect or its overwintering chamber, removal of the protective cover, elimination of food plants, and disruption of the insect life cycle generally killing many of the insects through direct contact, starvation or exposure to predators, and weather. The use of trap strip crops can control insect damage at the field edges and at the same time avail refuge and food for beneficial insects. Insect resistance is an important component of pest and disease management. Quality-based resistance can be induced in plants through management of nutrients and irrigation. Intercropping and biodiversity play an important role in pest management in organic farming.

Mechanical and Physical Pest Control

Devices that can be used to exclude insect pests from reaching crops in organic farming include, but not limited to, row covers, protective nets with varying mesh size according to the pest in question, and sticky paper collars that prevent crawling insects from climbing the trunks of trees. Water pressure sprays can be employed to dislodge insect pests such as aphids and mites from the plant surface. Insect vacuums, on the other hand, could be used to remove insects from plant surface and collect them into a collection box.

Biological Pest Control

Biological methods are the use of beneficial organisms that can be used in the field to reduce insect pest populations. Biological control is grouped into three categories: importation or classical biological control, which introduces pest's natural enemies to the locations where they do not occur naturally, augmentation involves the supplemental release of natural enemies, boosting the naturally occurring population, and conservation, which involves the conservation of existing natural enemies in the environment. The role of beneficial species on pests is of relatively greater importance in organic agriculture than in conventional agriculture, because organic growers do not have recourse to highly potent insecticides (such as synthetic pyrethroids) with which to tackle major pest problems.

Biopesticide Control

Biopesticides are characterized by having minimal or no risk to the environment, natural enemies, and nontarget organisms due to their mode of action, rapid degradation, and the small amounts applied to control pests. They are slow acting, have a relatively critical application times, and suppress rather than eliminate a pest population. Biopesticides have limited field persistence and shorter shelf life and present no residue problems. Thus, they are approved for pest management in organic crops.

Plant Protection Products (PPPs) Authorized in organic Farming

The crop protection in organic farming is holistic, and, hence, it is extremely difficult to separate inputs as plant nutrients (fertilizers) and plant protectants (pesticides). Plant protection products authorized for use in organic farming differ among countries depending on the differences in crops, pests, and cropping systems, as well as regulations and standards adopted by these countries. Organically approved pesticides fall into the following groups: biorational, inorganics, botanicals, microbial, oils, and soaps. The most widely used as insecticides are microorganisms, natural pyrethrins, rapeseed oil, and paraffin; the most widely used as fungicides are copper compounds, sulfur, and microorganisms. The rules of organic agriculture allow the use of unregistered products such as nettle slurry, which is used against aphids. It can be prepared on the farm or shared among farmers.

The basic substance concept was introduced by the EU regulation 1107 in 2009. It was defined as substance not intendedly used for plant protection purposes; however, it can still be used in protection of plants either directly or as a diluent. According to this definition, substances used as foodstuff such as vinegar and sunflower oil can be used as plant protection. The basic substances of plant and animal origin, which are used as foodstuff, can be legally used in crop protection in organic farming with the exception of being used as herbicides. These basic substances include chitosan hydrochloride, fructose, sucrose, Salix spp. cortex, and Equisetum arvense L. (field horsetail) which are used as elicitors of the plant self-defense mechanism. Sunflower oil, whey, and lecithins are used as fungicides, while vinegar is used as fungicide and bactericide, and Urtica sp. is used as insecticide, fungicide, and acaricide. In organic farming, only active substances listed in the Commission Regulation (EC) No. 889/2008 can be used. New update is frequently being made by the EC to add or remove PPPs from the list.

Table: Plant protection products approved by the European Union (EU) for use in organic farming.

Name of product	Purpose and specifications of use
Azadirachtin from the neem tree (*Azadirachta indica*).	
Beeswax	Used as protectant for treatment of cuts and wounds after pruning or in grafting.
Plant oils	Used for control of small-bodied insects such as thrips, aphids, and whiteflies.
Laminarin (from *Laminaria digitata*) or kelp or brown algae seaweed.	A polysaccharide from the group of the glucans, used to protect plants against fungi and bacteria. Kelp should be grown according to the organic standards.
Pheromones	Used only in traps and dispensers
Pyrethrins from the leaves of *Chrysanthemum cinerariaefolium*.	Used as insecticide
Pyrethroids (only deltamethrin or lambdacyhalothrin).	Used only in traps with attractants or pheromones.
Quassia from the plant *Quassia amara*.	Only insecticide and repellent.

Microorganisms, e.g., *Bacillus thuringiensis*, *Beauveria bassiana*, and *Metarhizium anisopliae*.	Origin should not be GMOs
Spinosad from the soil bacterium *Saccharopolyspora spinosa*.	Used as insecticide
Ethylene	Insecticidal fumigant against fruit flies
Paraffin oil	Used as insecticide against small-bodied insects
Fatty acids (soft soaps)	Insecticide against mite, thrips, and aphids
Lime sulfur (mixture of calcium hydroxide and sulfur).	Used as fungicide
Kieselgur (diatomaceous earth) from the hard-shelled diatom protist (chrysophytes).	Used as mechanical insecticide
Naturally occurring aluminum silicate (kaolin).	As insect repellent against a wide range of insects at a rate of 50 kg/ha.
Calcium hydroxide	Used as fungicide
Sodium hypochlorite (bleach or as javel water). It is a disinfectant with numerous uses, and its effect is due to the chlorine.	Used in seed treatment as viricide and bactericide.
Sulfur	Used as broad-spectrum inorganic contact fungicide and acaricide.
Copper compounds such as: copper hydroxide, copper oxychloride, copper oxide, tribasic copper sulfate, and Bordeaux mixture (copper sulfate and calcium hydroxide).	Used as fungicide and bactericide maximum of 6 kg copper per ha annually.
Sheep fat (obtained from fatty sheep tissues by heat extraction and mixed with water to obtain an oily water emulsion).	A triglyceride consisting predominantly of glycerine esters of palmitic acid, stearic acid, and oleic acid. A repellent by smell against vertebrate pests such as deer and other game animals. It should not be applied to the edible parts of the crop.
Quartz sand	Used as repellent against vertebrate pests

Impact of Pest Management in Organic Farming on the Environment

Pest management in organic farming depends mainly on crop husbandry and biological control. The prohibition of synthetic fertilizers and pesticides leads to conservation of natural enemies including predators and parasitoids. The absence of harmful pesticides also increases diversity of pollinators of crops and minimizes pesticide residues in food products. The community of microorganisms flourishes well in organically managed farms leading to increased organic matter decomposition, soil fertility, and sustainability of the ecosystem. Organic farming enhances the biodiversity of the ecosystem through multicropping and growing of hedges and refuges for beneficial insects as well as wildlife. Preserving biodiversity contributes much in reducing the initial invasion and subsequent establishment of organic farms by pests.

Organic Pest Management Approach

- Pests are indicators of how far a production system has strayed from the natural ecosystems it should imitate.

- Pests are attracted to a plant that is weak or inferior.

- In a well-balanced system, massive pest outbreaks are rare due to the presence of natural predators, parasites, and disease agents.

- Prophylactic, holistic approach vs. remedial approach.

- Not just treating symptoms.

- Pest problem usually indicates sub-optimal growing conditions and imbalance.

- Emphasis on biodiversity and optimal cultural practices.

The Disease Triangle

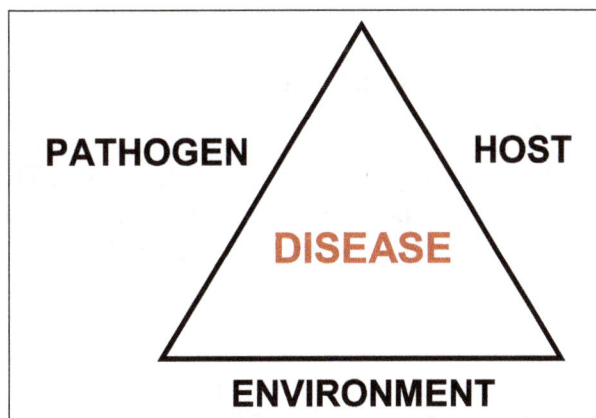

The disease triangle is a schematic that explains that in order to have disease you must have:

- The pathogen present on your farm. You can eliminate some pathogens through crop rotation, only using disease-free seed or plant stock, deep tillage to move soil pathogens deeper into the soil, etc.

- An appropriate host for the pathogen to live and reproduce. Crop rotation would really help here. One of the most important crop protection methods is using disease resistant varieties. Make sure you are matching the correct disease resistance to the problem on your farm. We hope convincing our to get your disease problems properly diagnosed.

- Suitable environmental conditions are needed for disease to develop. Pathogens require environmental cues, such as humidity, moisture and desirable temperatures to germinate, survive and infect. If these conditions are not met, pathogens cannot survive. Ex. downy mildews require a certain amount of free moisture present on the leaf surface to germinate and infect. If you can increase air flow and the drying of the leaf surface you can make life miserable for the pathogen.

Disease Causing Organisms

- Fungi.

- Bacteria.

Viruses.

Nematodes.

Symptoms

- Fungi cause spots, lesions, blights, yellowing of leaves, wilts, cankers, rots, fruiting bodies, mildews, molds, leaf spots, root rots, cankers, and blotches. Fungi are typically spread by wind, rain, soil, mechanical means and infected plant material.

- Bacteria cause water-soaking, spots, wilts, rots, blights, cankers, exudates, galls, yellowing, leaf spots, watery blotches, wilting. Bacteria are typically spread by rain, mechanical means, planting material, vectors (ex. bacterial wilt of cucurbits spread by cucumber beetle)

- Viruses cause mottling, leaf and stem distortions, mosaic patterns, rings and stunting. Viruses cause interesting symptoms, some are beautiful. Viruses are spread by mechanical means, vectors and in plant material.

- Nematodes cause wilting, stunting, yellowing of entire plants. This is because the roots of the plant are infected and the plant is starving or thirsty. Nematodes are spread by soil on equipment or workers boots or on infected plant material.

Control

The NOP has a hierarchical approach to pest management starting with System-based cultural practices then Mechanical and Physical Practices and finally Material-based (chemical, botanical, elemental) practices.

Cultural Control

Cultural control is your first line of defense:

- Promote healthy soils and healthy plants. Healthy soil is the hallmark of organic agriculture. An unhealthy plant is very attractive to diseases.

- Soils rich in organic matter are shown to increase soil biodiversity and help to create and abundance of beneficial soil microorganisms. Using compost has been shown to increase the suppressiveness of the soil by encouraging beneficial microorganisms, as well as inducing disease resistance in plants by simply having healthier plants.

- Exclusion:

  ○ Disease-free seeds, transplants or plant stock.

  ○ Prevent introduction of diseased plants or soil.

  ○ Disease free water source.

  ○ Control insects that can carry disease.

  ○ Soil solarisation.

- Disease resistant varieties.

- Good sanitation from the prior season. Remove diseased plants or weeds from the field. Don't put disease plants or weeds in the compost pile! Many diseases are resistant to high heats or can become resistant. Some pathogens form resistant structures that can tolerate unfavorable conditions.

- Always work infested fields last and clean off equipment.

- Disinfest tools.

- Plant on raised beds. Not only helps with avoidance of pathogen, but also good moisture drainage is key.

- Crop rotation:

  - 3 yrs between crops in the same family.

  - Some pathogens cause disease among multiple plant families.

- Plants adapted to area.

- Plant at proper depth (below crown or graft).

- Use only thoroughly composted material.

- Improve air circulation by staking, trellis or pruning.

- Water in the morning.

- Avoid overhead irrigation if possible.

Make Life Difficult for the Pathogen

Create an Unfavorable Environment for Pathogens

- Increase air movement

  - Trellising, high tunnels

- Increase soil drainage

- Avoid low-lying areas

- Row orientation

  - Maximize air movement

  - Minimize leaf wetness periods

- Irrigation management

- Drip Irrigation

- ◦ Mulches

- ◦ Plastic or plant-based

- Reduce splash dispersal of pathogens.

- Protect fruit from soilborne pathogens.

- Avoidance:

  - ◦ Plant your crop when disease isn't as big a problem.

  - ◦ Early blight and Cucurbit Downy Mildew.

Maximizing Disease Suppression with Compost

- Compost:

  - ◦ Cure 4 or more months.

  - ◦ Incorporate into soil several months before planting.

  - ◦ Inoculate with beneficial microorganisms, e.g. Trichoderma.

- Application:

  - ◦ 5-10 tons (dry weight)/A - rule of thumb.

  - ◦ Apply every year until significant organic matter improvement observed; watch for increases in P.

Produce Healthy Transplants

- Practice good sanitation in the greenhouse:

  - ◦ Use new or sanitized plug trays or flats and pathogen-free mixes.

  - ◦ Sanitize equipment.

  - ◦ Install solid flooring; raise seedling trays.

  - ◦ Limit movement of personnel and equipment between greenhouses.

  - ◦ Clean benches, greenhouse structure thoroughly after the crop; close up greenhouse.

Variety Selection

Disease tolerance is the ability of a plant to endure an infectious or non-infectious disease, adverse conditions or chemical injury without serious damage or yield loss

Disease resistance is when a plant possessed properties that prevent or impede disease development.

Pick varieties that are appropriate for your area.

Keep records of a cultivar's performance and the disease pressure each season.

Local heirlooms are generally better suited for a particular region. Use tissue culture plants (small fruits, some cut flowers, perennials) if available. These plants are often disease indexed. This is especially important for viruses. This is a specialized area and there are not always tissue culture plants available.

Physical/Mechanical Controls

Physical and Mechanical controls are very important in insect management on organic farms. They are also very important for disease control, especially in perennial cropping systems like fruit trees, small fruits and tree nurseries.

Some options for physical/mechanical management of plant disease include:

- Hand-picking
- Pruning
- Mulches
- Soil solarization

Pruning

- Prune out diseased plant parts.
- Increase light into canopy.
- Increase airflow.
- Helps spray penetrate all surfaces.
- Proper pruning for proper plant health.

Soil Solarisation

Used in greenhouses, seed beds, cold frames.

In greenhouses or in raised beds you can sterilize the soil or bench using heat produced by steam. You want to heat the coldest part of the soil to 82 °C for 30 minutes.

Soil solarization, a nonchemical technique, will control many soilborne pathogens and pests. This simple technique captures radiant heat energy from the sun, thereby causing physical, chemical, and biological changes in the soil. Transparent polyethylene plastic placed on moist soil during the hot summer months increases soil temperatures to levels lethal to many soilborne plant pathogens, weed seeds, and seedlings (including parasitic seed plants), nematodes, and some soil residing mites. Soil solarization also improves plant nutrition by increasing the availability of nitrogen and other essential nutrients.

Hot Water Seed Treatment

Research has shown that hot water seed treatment can help to decrease disease in seeds. Times and temps of seed treatment:

- Brussels sprouts, eggplant, spinach, cabbage, tomato = 122 F for 25 min.

- Broccoli, cauliflower, cucumber, carrot, collard, kale, kohlrabi, rutabaga, turnip = 122 F for 20 min.

- Mustard, cress, radish = 122 F for 15 min.

- Pepper = 125 F for 30 min.

- Lettuce, celery, celeriac – 118 F for 30 minutes.

It is important to note that cucurbit seeds can be damaged by hot water. Other cautions include:

- Use new, high quality seed.

- Treat a small sample first and test for germination.

- Treat close to time of planting (within weeks).

- Treat only once.

Material Control

Materials include:

- Elemental fungicides
  - o Copper and sulfur
- Biofungicides/Microorganisms
  - ◦ Ex. PlantShield, MycoStop, Companion
- Particle Film Barriers
  - ◦ Ex. kaolin clay
- Peroxides and Bicarbonates
- Compost Teas.

Sulfur

- Used effectively for powdery mildew on most crops.

- Labeled for rusts (grape and bean), botrytis (onions), black spot (rose).

- pH adjustment.

- Component of Bordeaux mixture.

- Lime sulfur - protectant dust or spray to control some fungal or bacterial diseases

- Helps control rust, powdery mildew (PM), brown rot.

Copper

- Controls some fungi and bacteria

  - Free Cu - Copper sulfate: Bordeaux mixture

  - Fixed Cu - Copper hydroxide, copper oxide, copper oxychloride, copper octanoate

  - Kocides - Restricted use, requires license, OMRI approved?

  - Safer Garden fungicide

  - Cu 12% or 0.4%: rust, scab, brown spot, black spot, others

- Nasty stuff, some certifiers won't let you use it.

Botanical/Horticultural Oils

Used successfully to control insects that spread disease. Especially viral diseases. Some are effective for fungi like powdery mildews and rust.

Biocarbonates and Peroxides

- Bicarbonates - Potassium Bicarbonate (baking soda)

  - Disrupts cell membrane K balance

  - PM Black spot, leaf spots, rusts for seed, transplants or established plants

  - Ex. Kaligreen

- Peroxides

  - Disinfest plant surface

  - Pre-plant, plant dip, foliar spray

  - Use on tools, trays, pots, surfaces

  - Ex. Oxi-Date.

Antibiotics

- Antibiotics -Streptomycin sulfate – Many brands for agricultural use to control bacteria, fireblight

  - Fertilome Fireblight spray : Also for bacterial wilt, stem rot, leaf spots and crown gall

  - Tetracycline – fireblight.

Biofungicides/Microorganisms

Antagonists/Competitors

- Trichoderma harzianum is the most researched
- Antifungal properties
    ◦ Bacillus spp.
- Plant growth aids
    ◦ Healthy roots, soil exploration
- Trichoderma
    ◦ Activate plant immune system
- Bacillus pumilus.

Compost Teas

Compost tea, in modern terminology, is a compost extract brewed with a microbial food source—molasses, kelp, rock dust, humic-fulvic acids. The compost-tea brewing technique, an aerobic process, extracts and grows populations of beneficial microorganisms

Pre-Plant Options

- Biofumigation
    ◦ Mustards, broccoli residue
    ◦ Muscodor
    ◦ Broad-spectrum activity
- Biocontrols – Contans, Advan LLC
- Coniothyrium minitans a fungi used pre-plant
- Narrow-spectrum (Sclerotinia only)
- Ex. lettuce drop, sclerotinia blight on peanut.

Scouting for Disease

- Why scout?
    ◦ Identify disease problems during the season
- To change practices this year
- For next year

- Scout your crops on a regular basis (calendar). Scouting supplies include:
  - Hand lens (10X or higher)
  - Paper for notes
  - Self sealing bags for samples
  - A marker or pen – field guide
  - Digital camera.

Using the Plant Disease and Insect Clinic

- Collect fresh, don't send over weekend.
- Send several examples.
- Crush proof container.
- Provide lots of information.
- Don't pull plants-do bag roots.
- Don't get soil on foliage.
- Press leaves.
- Contact lab.
- Use your extension agent for help.
- Collect as much information as possible.

Example:

Early blight (potato, tomato) Early blight is caused by two fungi (Alternaria solani and Alternaria tomatophila) that are a serious problem in tomatoes and potatoes but rarely effects peppers and eggplants. All of the above-ground portions of the plant can be affected throughout the growing season. The disease starts on the lower leaves with small circular spots that have a target appearance of concentric rings. Leaves develop yellow blighted areas and later the tomato fruit may rot on the stem end. Potato tubers can also become infected, but this is quite rare. The pathogen can overwinter in the soil on diseased plant residues.

Cultural Control

1. Use crop rotations of at least 3 years to non-hosts (away from tomato, potato and eggplant).

2. Provide optimum growing conditions and fertility. Stressed plants (including drought) are more susceptible to early blight.

3. Stake or cage plants to keep fruit and foliage away from soil.

4. Drip irrigation is preferred, or overhead irrigation starting before dawn, so that the plants are dry early in the day. The key is to keep the period of leaf wetness to a minimum.

5. Mulching helps to prevent splashing of spores from soil up to lower leaves.

6. Indeterminate tomato and late-maturing potato varieties are usually more resistant/tolerant to early blight.

7. Early blight can be seed-borne, so buy from a reliable supplier. Hot water seed treatment at 122 °F for 25 minutes is recommended to control early blight on tomato seed. See chlorine treatment procedures under bacterial diseases.

8. Disinfect stakes or cages with an approved product each season before using. Sodium hypochlorite at 0.5% (12x dilution of household bleach) is effective, and must be followed by rinsing, and proper disposal of solution. Hydrogen peroxide is also permitted.

Materials Approved for Organic Production

1. Copper products showed one good and one poor result in recent studies.

2. A Trichoderma harzanium product, Plant Shield HC, used as a drench at planting, showed fair to good results in NY state on tomatoes over three seasons.

3. Serenade, Bacillus subtilis. A protectant. Labeled, but considered only partially effective in UNH trails.

Downy Mildews

- Not true fungi:
    - Watermolds.
    - Swimming spores.
- Like cool wet weather.
- Overwinter as resistant spores in soil or infected plant material or blow in seasonally from diseased plantings.
- Effects vegetables and perennials, especially important in cucurbits, grapes and hops.
    - Important to note that downy mildews are very host specific. Ex. the downy mildew that infects cucurbits is not the same as the one that infects grapes.

Cultural Control

Resistant Varieties

- Planting date
    - We will get it in fall

- Avoid overhead irrigation.

- Forecasting site.

- If transplanting- make sure transplants are disease-free.

- Rotation important for overall plant health.

- High tunnels.

Materials for control of downy mildews:

- OMRI-listed products:

    - copper, neem oil, biofungicides (ex. Serenade or Sonata), peroxides (ex. OxiDate), and bicarbonates (ex. Kaligreen).

- Compost teas.

- Best option as an organic grower is to use a copper product:

    - Spray early in the morning to avoid phytotoxicity.

Spraying copper prior to disease development or at very early onset (very few, mild symptoms), may help suppress the disease, but will not offer 100% control under favorable conditions (cool, wet and humid weather).

Powdery Mildew

Effects many plants, but like downy mildew powdery mildews are host specific. Powdery mildews like it hot and relatively dry (humid but not wet). Powdery mildew is perhaps one of the easiest diseases to diagnose.

Cultural Control

- Resistant varieties.

- Plant in sunny areas with good air circulation.

- Avoid overhead irrigation.

- Avoid excess fertilization.

    - Slow release better.

- Remove infected plant material.

Materials for Powdery Mildews

- Sulfur is very effective.

- Kaligreen and Armicarb (potassium bicarbonate-baking soda); dilute solutions of hydrogen peroxide (Oxidate).

- ◦ These materials burn out the fungus growing on the surface, but do not provide protection against new infections; thus, repeated applications are important.

- Oils:

  - ◦ Saf-T-Side Spray Oil, Sunspray Ultra-Fine Spray Oil, or one of the plant-based oils such as neem oil or jojoba oil (e.g., E-rase).

  - ◦ Be careful some plants are sensitive, esp. when used in conjunction with sulfur.

- Dilutions of milk and whey (the dairy by-product) have been effective for controlling powdery mildew (Australia).

Biological Pest Control

Biological control, biocontrol, or biological pest control is a method of suppressing or controlling the population of undesirable insects, other animals, or plants by the introduction, encouragement, or artificial increase of their natural enemies to economically non–important levels. It is an important component of integrated pest management (IPM) programs.

The biological control of pests and weeds relies on predation, parasitism, herbivory, or other natural mechanisms. Therefore, it is the active manipulation of natural phenomena in serving human purpose, working harmoniously with nature. A successful story of biological control of pests refer to the human beings' capability to depict natural processes for their use and can be the most harmless, non–polluting, and self–perpetuating control method.

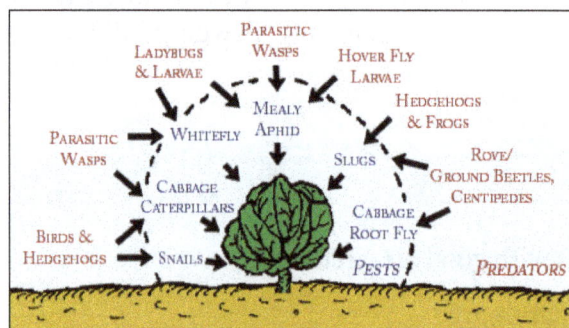

Illustration of the natural enemies of cabbage pests

In biological control, the reduction of pest populations is achieved by actively using natural enemies.

Natural enemies of the pests, also known as biological control agents, include predatory and parasitoidal insects, predatory vertebrates, nematode parasites, protozoan parasites, and fungal, bacterial, as well as viral pathogens. Biological control agents of plant diseases are most often referred to as antagonists. Biological control agents of weeds include herbivores and plant pathogens. Predators, such as lady beetles and lacewings, are mainly free–living species that consume a large number of prey during their lifetime. Parasitoids are species whose immature stage develops on or within a single insect host, ultimately killing the host. Most have a very narrow host range.

Many species of wasps and some flies are parasitoids. Pathogens are disease–causing organisms including bacteria, fungi, and viruses. They kill or debilitate their host and are relatively specific to certain pest or weed groups.

Strategies of Biological Control Methods

There are three basic types of biological control strategies; conservation biocontrol, classical biological control, and augmentative biological control (biopesticides).

Conservation Biocontrol

The conservation of existing natural enemies is probably the most important and readily available biological control practice available to homeowners and gardeners. Natural enemies occur in all areas, from the backyard garden to the commercial field. They are adapted to the local environment and to the target pest, and their conservation is generally simple and cost–effective. For example, snakes consume a lot or rodent and insect pests that can be damaging to agricultural crops or spread disease. Dragonflies are important consumers of mosquitoes.

Eggs, larvae, and pupae of Helicoverpa moths, the main insect pests of cotton, are all attacked by many beneficial insects and research can be conducted in identifying critical habitats, resources needed to maintain them, and ways of encouraging their activity. Lacewings, lady beetles, however fly larvae, and parasitized aphid mummies are almost always present in aphid colonies. Fungus–infected adult flies are often common following periods of high humidity. These naturally occurring biological controls are often susceptible to the same pesticides used to target their hosts. Preventing the accidental eradication of natural enemies is termed simple conservation.

Classical Biological Control

Classical biological control is the introduction of exotic natural enemies to a new locale where they did not originate or do not occur naturally. This is usually done by government authorities.

In many instances, the complex of natural enemies associated with an insect pest may be inadequate. This is especially evident when an insect pest is accidentally introduced into a new geographic area without its associated natural enemies. These introduced pests are referred to as exotic pests and comprise about 40 percent of the insect pests in the United States. Examples of introduced vegetable pests include the European corn borer, one of the most destructive insects in North America.

To obtain the needed natural enemies, scientists have utilized classical biological control. This is the practice of importing, and releasing for establishment, natural enemies to control an introduced (exotic) pest, although it is also practiced against native insect pests. The first step in the process is to determine the origin of the introduced pest and then collect appropriate natural enemies associated with the pest or closely related species. The natural enemy is then passed through a rigorous quarantine process, to ensure that no unwanted organisms (such as hyper parasitoids or parasites of the parasite) are introduced, then they are mass produced, and released. Follow–up studies are conducted to determine if the natural enemy becomes successfully established at the site of release, and to assess the long–term benefit of its presence.

There are many examples of successful classical biological control programs. One of the earliest successes was with the cottony cushion scale (Icerya purchasi), a pest that was devastating the California citrus industry in the late 1800s. A predatory insect, the Australian lady beetle or vedalia beetle (Rodolia cardinalis), and a parasitoid fly were introduced from Australia. Within a few years, the cottony cushion scale was completely controlled by these introduced natural enemies. Damage from the alfalfa weevil, a serious introduced pest of forage, was substantially reduced by the introduction of several natural enemies like imported ichnemonid parasitoid Bathyplectes curculionis. About twenty years after their introduction, the alfalfa area treated for alfalfa weevil in the northeastern United States was reduced by 75 percent. A small wasp, Trichogramma ostriniae, introduced from China to help control the European corn borer (Pyrausta nubilalis), is a recent example of a long history of classical biological control efforts for this major pest. Many classical biological control programs for insect pests and weeds are under way across the United States and Canada.

Classical biological control is long lasting and inexpensive. Other than the initial costs of collection, importation, and rearing, little expense is incurred. When a natural enemy is successfully established it rarely requires additional input and it continues to kill the pest with no direct help from humans and at no cost. Unfortunately, classical biological control does not always work. It is usually most effective against exotic pests and less so against native insect pests. The reasons for failure are often not known, but may include the release of too few individuals, poor adaptation of the natural enemy to environmental conditions at the release location, and lack of synchrony between the life cycle of the natural enemy and host pest.

Augmentative Biological Control

This third strategy of biological control method involves the supplemental release of natural enemies. Relatively few natural enemies may be released at a critical time of the season (inoculative release) or literally millions may be released (inundative release). Additionally, the cropping system may be modified to favor or augment the natural enemies. This latter practice is frequently referred to as habitat manipulation.

An example of inoculative release occurs in greenhouse production of several crops. Periodic releases of the parasitoid, Encarsia formosa, are used to control greenhouse whitefly, and the predaceous mite, Phytoseilus persimilis, is used for control of the two–spotted spider mite. The wasp Encarsia formosa lays its eggs in young whitefly "scales," turning them black as the parasite larvae pupates. Ideally it is introduced as soon as possible after the first adult whitefly are seen. It is most effective when dealing with low level infestations, giving protection over a long period of time. The predatory mite, Phytoseilus persimilis, is slightly larger than its prey and has an orange body. It develops from egg to adult twice as fast as the red spider mite and once established quickly overcomes infestation.

Lady beetles, lacewings, or parasitoids such as Trichogramma are frequently released in large numbers (inundative release) and are often known as biopesticides. Recommended release rates for Trichogramma in vegetable or field crops range from 5,000 to 200,000 per acre per week depending on level of pest infestation. Similarly, entomoparasitic nematodes are released at rates of millions and even billions per acre for control of certain soil-dwelling insect pests. Entomopathogenic fungus Metarhizium anisopliae var. acridum, which is specific to species of short–horned grasshoppers (Acridoidea and Pyrgomorphoidea) widely distributed in Africa, has been developed as inundative biological control agent.

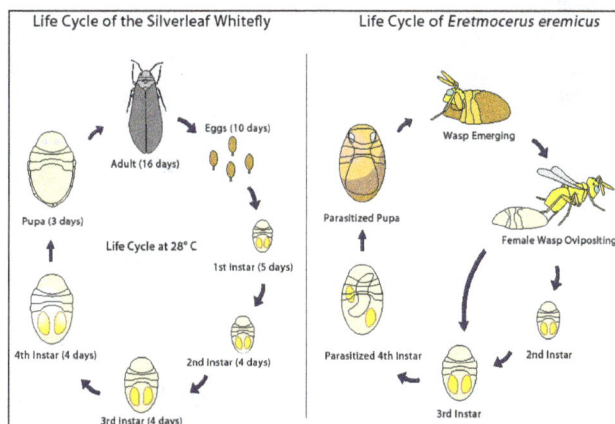

Life cycles of Greenhouse whitefly and its parasitoid wasp Encarsia Formosa

Habitat or environmental manipulation is another form of augmentation. This tactic involves altering the cropping system to augment or enhance the effectiveness of a natural enemy. Many adult parasitoids and predators benefit from sources of nectar and the protection provided by refuges such as hedgerows, cover crops, and weedy borders. Mixed plantings and the provision of flowering borders can increase the diversity of habitats and provide shelter and alternative food sources. They are easily incorporated into home gardens and even small-scale commercial plantings, but are more difficult to accommodate in large–scale crop production. There may also be some conflict with pest control for the large producer because of the difficulty of targeting the pest species and the use of refuges by the pest insects as well as natural enemies.

Examples of habitat manipulation include growing flowering plants (pollen and nectar sources) near crops to attract and maintain populations of natural enemies. For example, hover fly adults can be attracted to umbelliferous plants in bloom.

Biological control experts in California have demonstrated that planting prune trees in grape vineyards provides an improved overwintering habitat or refuge for a key grape pest parasitoid. The prune trees harbor an alternate host for the parasitoid, which could previously overwinter only at great distances from most vineyards. Caution should be used with this tactic because some plants attractive to natural enemies may also be hosts for certain plant diseases, especially plant viruses that could be vectored by insect pests to the crop. Although the tactic appears to hold much promise, only a few examples have been adequately researched and developed.

Different Types of Biological Control Agents

Predators

Ladybugs, and in particular their larvae which are active between May and July in the northern hemisphere, are voracious predators of aphids such as greenfly and blackfly, and will also consume mites, scale insects, and small caterpillars. The ladybug is a very familiar beetle with various colored markings, while its larvae are initially small and spidery, growing up to 17 millimeters (mm) long. The larvae have a tapering segmented gray/black body with orange/yellow markings nettles in the garden and by leaving hollow stems and some plant debris over–winter so that they can hibernate over winter.

Ladybird larva eating wooly apple aphids   Lacewings are available from biocontrol dealers

Hoverflies, resembling slightly darker bees or wasps, have characteristic hovering, darting flight patterns. There are over 100 species of hoverfly, whose larvae principally feed upon greenfly, one larva devouring up to 50 a day, or 1000 in its lifetime. They also eat fruit tree spider mites and small caterpillars. Adults feed on nectar and pollen, which they require for egg production. Eggs are minute (1 mm), pale yellow-white, and laid singly near greenfly colonies. Larvae are 8–17 mm long, disguised to resemble bird droppings; they are legless and have no distinct head. Therefore, they are semi–transparent with a range of colors from green, white, brown, and black. Hoverflies can be encouraged by growing attractant flowers such as the poached eggplant (Limnanthes douglasii), marigolds, or phacelia throughout the growing season.

Dragonflies are important predators of mosquitoes, both in the water, where the dragonfly naiads eat mosquito larvae, and in the air, where adult dragonflies capture and eat adult mosquitoes. Community–wide mosquito control programs that spray adult mosquitoes also kill dragonflies, thus removing an important biocontrol agent, and can actually increase mosquito populations in the long term.

Other useful garden predators include lacewings, pirate bugs, rove and ground beetles, aphid midge, centipedes, as well as larger fauna such as frogs, toads, lizards, hedgehogs, slow–worms, and birds. Cats and rat terriers kill field mice, rats, june bugs, and birds. Dogs chase away many types of pest animals. Dachshunds are bred specifically to fit inside tunnels underground to kill badgers.

Parasitoidal Insects

Most insect parasitoids are wasps or flies. For example, the parasitoid Gonatocerus ashmeadi (Hymenoptera: Mymaridae) has been introduced to control the glassy-winged sharpshooter Homalodisca vitripennis (Hemipterae: Cicadellidae) in French Polynesia and has successfully controlled about 95 percent of the pest density. Parasitiods comprise a diverse range of insects that lay their eggs on or in the body of an insect host, which is then used as a food for developing larvae. Parasitic wasps take much longer than predators to consume their victims, for if the larvae were to eat too fast they would run out of food before they became adults. Such parasites are very useful in the organic garden, for they are very efficient hunters, always at work searching for pest invaders. As adults, they require high–energy fuel as they fly from place to place, and feed upon nectar, pollen and sap, therefore planting plenty of flowering plants, particularly buckwheat, umbellifers, and composites will encourage their presence.

Four of the most important groups are:

- Ichneumonid wasps: (5–10 mm) Prey mainly on caterpillars of butterflies and moths.

- Braconid wasps: Tiny wasps (up to 5 mm) attack caterpillars and a wide range of other insects including greenfly. It is a common parasite of the cabbage white caterpillar, seen as clusters of sulphur yellow cocoons bursting from collapsed caterpillar skin.

- Chalcid wasps: Among the smallest of insects (<3 mm). It parasitizes eggs/larvae of greenfly, whitefly, cabbage caterpillars, scale insects, and strawberry tortrix moth.

- Tachinid flies: Parasitize a wide range of insects including caterpillars, adult and larval beetles, true bugs, and others.

Parasitic Nematodes

Nine families of nematodes (Allantone-matidae, Diplogasteridae, Heterorhabditidae, Mermithidae, Neotylenchidae, Rhabditidae, Sphaerulariidae, Steinernematidae, and Tetradonematidae) include species that attack insects and kill or sterilize them, or alter their development. In addition to insects, nematodes can parasitize spiders, leeches, annelids, crustaceans and mollusks. An excellent example of a situation in which a nematode may replace chemicals for control of an insect is the black vine weevil, Otiorhynchus sulcatus, in cranberries. Uses of chemical insecticides on cranberry either are restricted or have not provided adequate control of black vine weevil larvae. Heterorhabditis bacteriophora NC strain was applied, and it provided more than 70 percent control soon after treatment and was still providing that same level of control a year later.

Many nematode–based products are currently available. They are formulated from various species of Steinernema and Heterorhabditis. Some of the products found in various countries are ORTHO Bio–Safe, BioVector, Sanoplant, Boden-Ntitzlinge, Helix, Otinem, Nemasys, and so forth. A fairly recent development in the control of slugs is the introduction of "Nemaslug," a microscopic nematode (Phasmarhabditis hermaphrodita) that will seek out and parasitize slugs, reproducing inside them and killing them. The nematode is applied by watering onto moist soil, and gives protection for up to six weeks in optimum conditions, though is mainly effective with small and young slugs under the soil surface.

Plants to Regulate Insect Pests

Choosing a diverse range of plants for the garden can help to regulate pests in a variety of ways, including:

- Masking the crop plants from pests, depending on the proximity of the companion or intercrop.

- Producing olfactory inhibitors, odors that confuse and deter pests.

- Acting as trap plants by providing an alluring food that entices pests away from crops.

- Serving as nursery plants, providing breeding grounds for beneficial insects.

- • Providing an alternative habitat, usually in a form of a shelterbelt, hedgerow, or beetle bank, where beneficial insects can live and reproduce. Nectar–rich plants that bloom for long periods are especially good, as many beneficials are nectivorous during the adult stage, but parasitic or predatory as larvae. A good example of this is the soldier beetle, which is frequently found on flowers as an adult, but whose larvae eat aphids, caterpillars, grasshopper eggs, and other beetles.

The following are plants often used in vegetable gardens to deter insects:

Plant	Pests
Basil	Repels flies and mosquitoes.
Catnip	Deters flea beetle.
Garlic	Deters Japanese beetle.
Horseradish	Deters potato bugs.
Marigold	The workhorse of pest deterrents. Discourages Mexican bean beetles, nematodes and others.
Mint	Deters white cabbage moth, ants.
Nasturtium	Deters aphids, squash bugs and striped pumpkin beetles.
Pot Marigold	Deters asparagus beetles, tomato worm, and general garden pests.
Peppermint	Repels the white cabbage butterfly.
Rosemary	Deters cabbage moth, bean beetles and carrot fly.
Sage	Deters cabbage moth and carrot fly.
Southernwood	Deters cabbage moth.
Summer Savory	Deters bean beetles.
Tansy	Deters flying insects, Japanese beetles, striped cucumber beetles, squash bugs and ants.
Thyme	Deters cabbage worm.
Wormwood	Deters animals from garden.

Pathogens to be used as Biopesticides

Various bacterial species are widely used in controlling the pests as well as weeds. The best–known bacterial biological control which can be introduced in order to control butterfly caterpillars is Bacillus thuringiensis, popularly called Bt. This is available in sachets of dried spores, which are mixed with water and sprayed onto vulnerable plants such as brassicas and fruit trees. After ingestion of the bacterial preparation, the endotoxin liberated and activated in the midgut will kill the caterpillars, but leave other insects unharmed. There are strains of Bt that are effective against other insect larvae. Bt. israelensis is effective against mosquito larvae and some midges.

Viruses most frequently considered for the control of insects (usually sawflies and Lepidoptera) are the occluded viruses, namely NPV, cytoplasmic polyhedrosis (CPV), granulosis (GV), and entomopox viruses (EPN). They do not infect vertebrates, non–arthropod invertebrates, microorganisms, and plants. The commercial use of virus insecticides has been limited by their high specificity and slow action.

Fungi are pathogenic agents to various organisms including the pests and the weeds. This feature is intensively used in biocontrol. The entomopathogenic fungi, like Metarhizium anisopliae,

Beauveria bassiana, and so forth cause death to the host by the secretion of toxins. A biological control being developed for use in the treatment of plant disease is the fungus Trichoderma viride. This has been used against Dutch Elm disease, and to treat the spread of fungal and bacterial growth on tree wounds. It may also have potential as a means of combating silver leaf disease.

Significance of Biological Control

Biological control proves to be very successful economically, and even when the method has been less successful, it still produces a benefit–to–cost ratio of 11:1. The benefit–to–cost ratios for several successful biological controls have been found to range from 1:1 to 250:1. Further, net economic advantage for biological control without scouting vs. conventional insecticide control ranged from $ 7.43 to $ 0.12 per hectare in some places. It means that even if the yield produce under biological control be below that for insecticidal control by as much as 29.3 kilos per hectare, the biological control would not lose its economic advantage.

Biological control agents are non–polluting and thus environmentally safe and acceptable. Usually they are species specific to targeted pest and weeds. The biological control discourages the use of environmentally and ecologically unsuitable chemicals, so it always leads to the establishment of natural balance. The problems of increased resistance in the pest will not arise, as both biological control agents and the pests are in complex race of evolutionary dynamism. Because of chemical resistance developed by the Colorado potato beetle (CPB), its control has been achieved by the use of bugs and beetles (Hein).

Negative Results of Biological Control

Biological control tends to be naturally self–regulating, but as ecosystems are so complex, it is difficult to predict all the consequences of introducing a biological controlling agent. In some cases, biological pest control can have unforeseen negative results, that could outweigh all benefits. For example, when the mongoose was introduced to Hawaii in order to control the rat population, it predated on the endemic birds of Hawaii, especially their eggs, more often than it ate the rats. Similarly, the introduction of the cane toad to Australia 50 years ago to eradicate a beetle that was destroying sugar beet has been spreading as a pest throughout eastern and northern Australia at a rate of 35 km/22 mi a year. Since the cane toad is poisonous, it has few Australian predators to control its population.

Biological Control of Insect Pests

Biological control is the use of living organisms to maintain pest populations below damaging levels. Natural enemies of arthropods fall into three major categories: predators, parasitoids, and pathogens.

Predators

Predators catch and eat their prey. Some common predatory arthropods include ladybird beetles, carabid (ground) beetles, staphylinid (rove) beetles, syrphid (hover) flies, lacewings, minute pirate bugs, nabid bugs, big-eyed bugs, and spiders.

Figure: Preying mantid consuming insect prey

Parasitoids

Parasitoids (sometimes called parasites) do not usually eat their hosts directly. Adult parasitoids lay their eggs in, on, or near their host insect. When the eggs hatch, the immature parasitoids use the host as food. Many parasitoids are very small wasps and are not easily noticed. Tachinid flies are another group of parasitoids. They look like large houseflies and deposit their white, oval eggs on the backs of caterpillars and other pests. The eggs hatch, enter the host, and kill it. Parasitoids often require a source of food in addition to their host insect, such as nectar or pollen.

Figure: Parasitic ichneumonid wasp

Pathogens

Pathogens are disease-causing organisms. Just as many other organisms get sick, so do insects. The main groups of insect disease-causing organisms are insect-parasitic bacteria, fungi, protozoa, viruses, and nematodes. Biological control using pathogens is often called microbial control. One very well-known microbial control agent that is available commercially is the bacterium Bacillus thuringiensis (Bt). Because not all formulations of Bt are approved for use in organic systems, it is important to check with your certifier before using this. Several insect-pathogenic fungi are used as microbial control agents, including Beauveria, Metarhizium, and Paecilomyces. These are most often used against foliar insect pests in greenhouses or other locations where humidity is relatively high. Nuclear polyhedrosis viruses (NPV) and granulosis viruses (GV) viruses are available to control some caterpillar pests. The insect-parasitic (entomopathogenic or insecticidal) nematodes, Steinernema and Heterorhabditis, infect soil-dwelling insects and occur naturally or can be purchased. As with all biological control agents, it is especially important to match the correct microbial control agent with the correct pest in order for them to be effective.

Figure: Uninfected Beet armyworm (bottom), and beet armyworm killed by a nuclear polyhedrosis virus (NPV)

Approaches to Biological Control

Biological control can be natural: conservation of natural enemies or applied: inoculation or inundation.

Conservation of Natural Enemies

In many cases, purchasing natural enemies to provide biological control agents is not necessary. Natural enemies are common and a grower can design production systems to attract and keep the natural enemies in the system by providing environmental conditions conducive to the enemies' survival. Farmscaping is a term sometimes used to describe the creation of habitat to enhance the chances for survival and reproduction of beneficial organisms. For example, many adult predators and parasitoids feed on nectar and pollen, so it is essential to have these resources nearby. Having several species of pollen- and nectar-producing plants in an area will provide resources more continuously than only having one species. Many members of the Apiaceae (also known as Umbelliferae) family are excellent insectary plants. The flowers of fennel, coriander (cilantro), dill, and wild carrot are especially attractive to parasitoid wasps.

Organic mulches and crop residue moderate fluctuations of temperature and moisture and can provide hiding places for soil-dwelling insect predators such as ground (carabid) and rove (staphylinid) beetles, spiders, and centipedes. Undisturbed areas, such as windbreaks, hedgerow, or strips of perennial vegetation within fields (beetle banks), provide refuge habitat where beneficial insects can live and reproduce. Other habitats provided by farmscaping include water, alternate prey, perching sites, overwintering sites, and wind protection. Some refuge planting can harbor pests, so the success of farmscaping efforts depends on knowledge of pests and beneficial organisms.

Good soil management that returns organic matter to the soil to support an active food web can support vigorous plant growth and conditions that favor soil dwelling natural enemies, e.g., ground beetles. However, high organic matter and abundant crop residues can favor some pests, such as slugs, cutworms, wireworms, and root maggots.

Even pesticides allowed in organic production are insecticidal, and beneficial insects are often susceptible to the same pesticides used to control pest insects. If a pesticide must be used to control a pest outbreak, it should be applied in a manner to conserve beneficial insects. Application methods that result in low environmental exposure of beneficial organisms to pesticides should be used—for example, enclosed baits, low volume or spot treatment, or application at times of day when beneficials are not active.

Inoculation and Inundation

Inoculation and inundation involve the supplemental release of natural enemies to build populations of beneficial organisms. Many biological and microbial control agents are commercially available for purchase.

An inoculative approach involves the release of natural enemies at a critical time of the season to augment natural populations already present, but in numbers too low for effective pest management.

An inundative approach involves the application of a large number of organisms much in same manner as a pesticide. The applied organisms, which may or may not become established, can be used for relatively fast-acting, short-term control. Parasitoids such as Trichogramma are often released in vegetable or field crops at a rate of 5,000 to 200,000 per acre per week depending on level of pest infestation. Insect-parasitic (entomopathogenic) nematodes are often applied at a rate of 1 million to 1 billion nematodes per acre.

Microbial Control

Microbial control of insects is achieved through the inundative application of allowable formulations of insect-pathogenic bacteria (e.g., Bacillus thuringiensis), insect-pathogenic fungi (e.g., Beauveria bassiana), or insect viruses.

Information about rates and timing of release are available from suppliers of beneficial organisms. The quality of commercially available biocontrol agents is an important consideration. Biological and microbial control agents are living organisms, and must not be mishandled during shipping, storage, or application.

Organic Weed Control

Weeds compete with crops for nutrients, light and water. This is why it is important to keep them in check. Growing organically allows you to manage weeds creatively and effectively, by careful planning of your planting and using mulches. Not all weeds are bad - some attract pollinators and some can improve the soil. And of course, most weeds can be composted, which in turn will add nutrients to your soil.

Five Things about glyphosate:

- It is a toxic herbicide used to kill weeds.

- On its own, glyphosate has limited toxicity. However, it is commonly mixed in chemical formulations to have maximum effect. These formulations, such as Roundup or Weedol Path Clear, are potentially far more dangerous.

- The European Food Safety Authority (EFSA) says that glyphosate is safe. However, most of their research is industry led, and they haven't tested the various individual commercial formulations.

- Independent research indicates that glyphosate is not only probably carcinogenic, but that it also affects the body's endocrine system – causing problems in the liver and kidneys. Industry testers dispute this, but have declined to reveal all the results of their safety tests.

- Over 30% of the bread in the UK contains traces of glyphosate. While not necessarily toxic in small amounts, this gradual and persistent intake could create a health risk.

The rid of weeds without toxic chemicals (Roundup, Weedol Path Clear, Resolva etc):

- Patios/paths and other hard surfaces.

- Use boiling water, or a sharp knife or trowel.

- Flame or thermal weeders are effective, especially on young weeds, and relatively cheap to buy.

- Organic weedkiller is usually based on pelargonic acid - which kills the foliage, but doesn't penetrate the root.

- Vinegar is not recommended. Some local councils use DEFRA approved acetic acid (the basis of vinegar). As this acid is a chemical compound it is not organic. Domestic vinegar from a bottle has not been proved effective. Bleach is also not recommended.

- When creating a new path or patio, make an impenetrable foundation layer: a geotextile membrane or a substantial mix of hardcore rubble and sand, firmly flattened to exclude all light and reduce moisture.

- Cracks between pavers should be filled with mortar - not sand, which provides the ideal medium for weeds to germinate. Lime mortars are more environmentally friendly than cement mixes.

Larger areas, such as an overgrown allotment or veg patch:

- To clear an overgrown growing area, slash down the high standing weeds and then cover with a thick compost manure mulch (at least 20 cms) and a plastic sheet. Without light the weeds will weaken and eventually die off. Use the slashed foliage and stems on the compost heap.

- Dig up deep rooted weeds, such as dandelions and docks. Put foliage on the compost heap, and drown the roots in a bucket of water for a month or so. The water can be used as a liquid feed.

- Persistent weeds such as bindweed and ground elder have to be dug over regularly, removing as much root as possible. Every little bit. In some cases, it is worth digging up individual plants which you want to keep, cleaning their roots of the weed's root fragments, and then replacing in the bed which has also been dug over. Persistent digging, removing roots, will eventually – perhaps over a couple of years – weaken the plant and make it easier to keep on top of. Put foliage and roots (but not flowers or seed heads) into a black plastic sack. Tie up the top and leave in an out-of-the-way corner until it turns into gooey sludge, then compost it.

Remember, it is not just your own health that will benefit from not using toxic chemicals – you will be helping other life forms to thrive in your growing area. Leaving some weeds, such as a discreet area of stinging nettles, will provide food for pollinating insects, as well as leaves to make a liquid feed, high in nitrates.

Direct Weed Control

Although cultural methods provide the basis for weed management in organic crop rotations it is likely that some form of direct action will be needed against weeds to prevent crop loss at some time. Before taking any action it is important to take an overview and assess whether the weeds present are likely to develop to such an extent that they will cause an immediate loss of crop or will store up potential future problems (e.g. by shedding seed and adding to the soil seed bank so exacerbating future weed problems). If the weed burden is judged to have the potential to cause damage the cost of this should be offset against the likely costs of any immediate or future direct control measures so that direct weeding is only underaken when it is economically beneficial to do so.

- Mechanical weed control provides an overview of the range of options and implements available for direct mechanical weed control in the field.

- Manual weed control is still an important component of many weed management programmes.

- Thermal weed control is becoming more popular.

- Mulching provides a physical barrier to weed development and is often used in horticultural crops to control weeds.

- Biological weed control aims to get insects, pathogens or even other plants to do the work of weed management for the farmer.

- Allelopathy can be regarded as a component of biological control in which plants are used to reduce the vigour and development of other plants.

Mechanical Weed Control

Mechanical weed control may involve weeding the whole crop, or it may be limited to selective inter-row or intra-row weeding. Machines can be used to kill weeds by burying, cutting or uprooting. Tools without a cutting action are only effective on small weeds. Inter-row implements have been designed that control weeds within the crop row by directing soil along the row to cover small weeds. With slow germinating crops inter-row weeding may have to be delayed until the crop seedlings emerge. In some situations it may be possible to include a few seeds of a fast germinating crop in the seed mixture to give an early indication of the position of the crop row.

Mechanical weeders range from basic hand tools to sophisticated tractor driven or self-propelled devices. These may include cultivating tools such as hoes, harrows, tines and brush weeders, cutting tools like mowers and strimmers, as well as implements like thistle-bars that may do both. Two wheeled pedestrian or walking tractors are a smaller alternative that can power a similar range of implements. Custom-made basket or cage-wheeled weeders, with gangs of rolling wire cylinders, offer another way to deal with seedling weeds in a friable soil. The choice of implement, and the timing and frequency of its use may depend on the crop and on the weed population. Some implements, such as fixed harrows, are thought more suitable for arable crops, while others like inter-row brush weeders may be considered to be more effective for horticultural use.

The weather and soil conditions under which the operation is carried out will have a major influence on its efficacy. Soil type, surface structure and moisture content affect the choice and efficacy of mechanical weed control implements. The options may be more limited on heavy or stony soils, most implements work better on light, stone-free soils. Mechanical weeding is less effective when soils are wet during or after weeding operations. Some implements like brushweeders are able to work at a higher soil moisture content than others. Conversely, brushweeders do not work well on dry soils where the surface has capped because the brushes cannot penetrate the crust.

The optimum timing for mechanical weed control is influenced by the competitive ability of the crop. A single inter-row cultivation at any time may provide excellent weed control in a crop like transplanted broccoli that rapidly develops a broad, shading leaf canopy but may be poorer in crops like sweet corn (Zea mays) where early growth is slow, or in green beans (Phaseolus vulgaris) where the growing season is relatively long. In the UK, the optimum timings for mechanical weed control have been defined for onions and for carrots grown in both organic and conventional systems. In organic winter cereals studies have found that, found that corn poppy (Papaver rhoeas) was more effectively controlled in the autumn whilst chickweed (S. media) was controlled best in spring when using a spring-tine weeder.

Weed morphology and stage of growth will also influence the selection and efficacy of weeding implement. In experiments to determine the type of physical damage that gave the most effective control of a range of seedling weeds it has been found that burial to 1 cm depth is the most effective treatment, closely followed by cutting at the soil surface. Plants need to be buried totally to be killed but plant size, angle and growth habit influence the depth of covering required. Some advisors suggest that if weeds have emerged you are already late with your weeding operation, and that the best time to kill them is at the white thread stage. At high weed densities, even with the most effective mechanical weeders, sufficient weeds are likely to survive control measures and profoundly reduce crop yield in cereals and direct control needs to be linked with long term preventative measures to maintain the weed population at a manageable level.

With most mechanical weeding implements, operator skill, experience and knowledge are critical to success. Drawbacks to mechanical weed control include low work rates, delays due to wet conditions, and the subsequent risk of weed control failure as weeds become larger. A review of the merits of six different mechanical weeding mechanisms in controlling inter-row weeds at different growth stages and at different tractor speeds indicated that weed control was not necessarily better at earlier weed stages and weeding too early often missed late germinating weeds. Increasing forward speed did not improve the performance of all the implements equally.

There may be some disadvantages to the greater use of mechanical weed control. The additional cultivations associated with mechanical weeding could harm soil structure and possibly encourage soil erosion. The increased mineralisation of soil nitrogen due to cultivation may be seen by some growers as a problem and by others as an advantage (although this is likely to be limited in effect). There is concern about the impact of mechanical weeding on ground nesting birds and management practices may some alteration to minimise disruption at critical times although evidence is at times contradictory. Any soil cultivation will also contribute to the movement of weed seeds in the soil. Studies of the horizontal movement of freshly shed seeds have shown that a sequence of cultivations could move seeds over 2 m horizontally.

Manual Weed Control

Manual methods of weed control are still widely used in organic systems. Hand weeding is most useful on annual weeds and some perennial weeds. There are times when hand roguing individual plants or patches of plants is the most effective way of preventing them spreading and multiplying. It is widely used for dealing with the removal of difficult-to-control species such as docks, thistles and ragwort.

Manual methods of weed control are also widely used in intensive horticultural crops where it is important to perform a good first weed to prevent weed competition. Hand weeding can often follow after a mechanical inter-row weeding operation in order to thoroughly remove weeds in the crop row. It is a practical method of removing weeds within rows and hills where a cultivating implement cannot be used. It obviously requires more labour than other direct weed control methods and therefore costs are likely to be higher so it is only employed by growers with high value crops like vegetables.

It is generally more efficient for groups or gangs of workers to hoe or hand weed crops as a team, whether directly pulling the weeds or using some type of hoe. Hand rogueing or pulling is a widely used technique for patches of weeds or removing. There are many modern hoe designs that are more comfortable to use than traditional designs and these should be investigated where large areas are being covered. Some designs include the stirrup hoe, the diamond hoe and the collinear hoe as well as wheeled push hoes. Other tools have been designed to tackle specific weeds such as docks, thistles or ragwort.

In more mechanised systems teams of workers lie on a flat-bed weeder pulled by a tractor or on other specially designed machines. The speed of the machines can be adjusted to accommodate the level of weeds in the crop.

Thermal Weed Control

One of the earliest forms of thermal weed control, stubble burning, is now banned because of the smoke and other hazards it created. However, this traditional form of thermal weed control was effective in reducing the number of viable weed seeds returned to the soil after cereal harvest. Soil surface temperatures under the burning straw reached in excess of 200 °C for 10 -30 seconds and reduced the viability of freshly shed wild oat (Avena fatua) and blackgrass (Alopecurus myosuroides) seed by up to 30% and 80% respectively. Current thermal weed control methods use a variety of thermal weeders to generate the heat needed to kill weed seeds and weed seedlings.

Flaming equipment has been developed in several countries including Germany, Holland, Sweden and Denmark, and a range of tractor and smaller hand operated burners is available in the UK. The main fuel used in the burners is liquefied petroleum gas (LPG) usually propane.

Mulching

Covering or mulching the soil surface can reduce weed problems by preventing weed seed germination or by suppressing the growth of emerging seedlings. Mulches are generally ineffective against established perennial weeds. A mulch may take many forms: a living plant ground cover, loose particles of organic or inorganic matter spread over soil, and sheets of artificial or natural materials laid on the soil surface. Residues from preceding crops may be used to form a mulch but this is discussed in more detail in the use of cover crops to suppress weeds. With mulches consisting of organic materials, crop stand and vigour, particularly of direct-seeded small-seeded crops, may be reduced by chemicals released from the decomposing residues.

It is most practical to use mulches in well-spaced crops, particularly transplants. Plastic sheeting and straw mulches have long been used in soft fruit such as strawberries. In perennial crops and some other situations mulches may be intended to remain effective for many years. Mypex, a black, woven, polypropylene mulch, is expected to last for up to three crops (9-10 years). These mulches may be expensive but labour costs are reduced in the long term. Other uses for mulches include as an alternative to cultivation to clear vegetation before cropping by leaving them in place for 12 to 18 months. In freshly prepared seedbeds, short term mulching can be used to manipulate or reduce weed seedling emergence, by for instance, laying black plastic on the seedbed 2 to 8 weeks and then lifting it before planting brassicas or other crops.

The high cost of mulching makes it economic only for high value horticultural crops unless there is another reason for its use. In addition to weed control, mulches may be used: to prevent soil erosion, reduce pest problems, to aid moisture and to prevent nitrate loss. In strawberries, rain splash dispersal of disease spores like those of black spot (Colletotrichum acutatum) is reduced by straw mulch. Mulches can also moderate soil temperatures. Organic mulches in particular reduce heat loss from the soil in cold conditions and help to prevent frost heave. In hot weather the mulch slows down the warming of soil.

Types of Mulches

- Living mulches: consist of a dense stand of low growing plants established prior to or after the crop. The undersowing of cereals with clover and grass could be seen as forming a living mulch. It has been argued that annual weeds would provide a natural ground cover if managed properly. Living mulches are sometimes referred to as cover crops, but they grow at least part of the time simultaneously with the crop. Cover crops are generally killed off prior to crop establishment.

Often, the primary purpose of a living mulch is that of improving soil structure, aiding nutrition or avoiding pest attack, and weed suppression may be just an added benefit. Maintaining vegetation cover is important for preventing soil erosion, nitrate leaching and weed emergence in slowly developing crops like maize. An investigation of the influence of different mulch species on weed density and diversity indicated that weed numbers were reduced and maize yield was not affected where growth of the mulch was reduced by cutting or flaming treatments. When the growth of a living mulch is not restricted, or when soil moisture is inadequate, even a relatively vigorous crop like potato may suffer competition and loss of yield. Studies have been made of the use of living mulches to suppress weed emergence in horticultural crops but there are many different factors to take into account and it is difficult to get the balance between crop and mulch right. Living mulches are well suited to use in perennial crops such as fruit where self-reseeding is an advantage. However, even in established apple and apricot orchards a living mulch growing along the planted row may depress crop growth. In amenity situations, ground covering plants are established to form a dense canopy and suppress weed germination and growth.

- Particle mulches: May be organic or inorganic. Loose materials like straw, bark and composted municipal green waste provide effective weed control but the depth of mulch needed to suppress weed emergence is likely to make transport costs prohibitive unless the material is produced on the farm. It has been shown that a 3 cm layer of compost was needed to prevent the emergence of annual weeds and weed control usually improves as the thickness of the organic mulch increases. Weed seeds in the mulch itself can be a problem if the composting process has not been fully effective or there is contamination by wind blown seeds. In straw mulches, volunteer cereal seedlings are a particular problem due to shed cereal grains and even whole ears remaining in the straw after crop harvest. With particle mulches like straw that consist of light materials there is the possibility of them being blown around by the wind. Organic mulches like straw with a high carbon to nitrogen ratio may deplete the soil of nitrogen as they decompose. Mulch improves water filtration into the soil and prevents the compaction and erosion that heavy rainfall can cause.

Before applying a particle mulch weeds should be removed, dry soil should be moistened, and compacted soil loosened. Old mulch should be removed or incorporated to prevent a build-up. Most mulch is applied 7.5 to 10 cm deep. Coarser textured materials require thicker layers. On sandy soils, the mulch layer needs to be deeper than on heavy or wet soils. The mulch should be raked periodically and topped up if necessary. Machinery has been developed for applying/spreading particle mulches. Bark blowers are widely available. Self-feeding straw blowers can handle 1-2 bales per minute and can cover an acre per hour. Flail-beater chains break the straw into 5-10 cm lengths. For green waste there are all-in-one collectors, shredders and spreaders.

- Sheeted mulches: A layer of material such as plastic, paper of woven fabric covers the soil surface. Black polyethylene mulches are widely used for weed control in organic and conventional systems in the UK and elsewhere. Clear mulches are better than black for warming the soil but do not control the weeds. Plastic mulches have been developed that selectively filter out the photosynthetically active radiation (PAR) but let through infra red light to warm the soil. Infra- red transmitting (IRT) mulches have been shown to be effective in controlling weeds. Various colours of woven and solid film plastics have been tested in the field. White and green coverings had little effect on the weeds, brown, black, blue, and white on black (double colour) films prevented weeds emerging. There are indications that mulching films, like white on black, with a higher rate of light reflectance are beneficial to the crop. Light reflectance may also affect the behaviour of certain insects, and plastic mulches in a greater array of colours are likely to become available. The woven and non-woven polypropylene films or geotextiles (like Mypex) are sometimes referred to as weed barriers and landscape fabrics. They are more durable than polyethylene films permitting multi-year use and are permeable to water. There are advantages both in reduced laying and disposal costs compared with single season materials.

Sheeted materials are relatively expensive and are usually laid by machine. Machinery has been developed that will raise the soil into beds and lay the plastic mulch, securing it at the edges. Beds can be prepared in advance of crop planting. Heavy duty plastic is used for long term crops such as perennial herbs. Woven polypropylene fabrics allow water to penetrate and are less likely to scorch crops when temperatures are high. Non-woven black fabric mulch may not be sufficiently opaque to prevent weed growth completely. After cropping, lifting and disposal may be a problem with plastic and other durable mulches and this adds to the overall costs. Even the degradable plastics may break into fragments that litter the soil. Sheeting made from paper and other natural fibres have the advantage of breaking down naturally, and can be incorporated into the soil after use. Paper mulches have compared favourably with black polyethylene in trials with transplanted lettuce, Chinese cabbage and calabrese in the UK although tearing and wind blowing can be a problem.

Biodegradable film mulches have been developed that adjust their degradation according to the soil and weather conditions. The polymers are strong enough to withstand mechanical laying.

Biological Weed Control

Biological weed control involves the release of organisms that attack plants to control weeds. The aim of biological control is to shift the balance of competition between the weed and the crop in favour of the crop and against the weed. The biological control agent, normally a fungus or insect, may not necessarily kill the target weed but should, at the least, reduce its vigour and competitive ability. From a practical point of view the organism or agent should prevent the weed setting seed or producing other reproductive parts. There is considerable potential for encouraging the use of native biological control agents against weeds and substantial research effort has been put into biological control in general. However, the application of biological weed control in agricultural systems in Europe has proved difficult and their are no well documented successes.

In practice, there are three basic types of biological control:

1. Classical (or innoculative) biological control involves the release of exotic natural enemies to control exotic weeds and has been successful against weeds like thistles in the US and Australia where weevils (native to Europe) have been introduced onto the thistles. It has been suggested that some introduced weeds like hogweed (Heracleum mantegazzium), Himalayan balsam (Impatiens glandulifera) and the Japanese knotweeds (Reynoutria spp.) would be ideal candidates for classical biological control but so far it has only been attempted with bracken (Pteridium aquilinum) where attempts to use two South African moths as potential biological control agents were not successful.

The introduction of a classical biological control agent may not be deliberate. A rust, Puccinia lagenophorae, of Australian origin, which attacks a range of Senecio species, was unknown in Europe before 1960 but has since become established in France and the UK on groundsel where it reduces the viability of groundsel plants on which it can be regularly found. It is unknown how this pathogen reached Europe or how it established.

2. Inundative control involves the mass production and release of native natural enemies against native weeds, for example rust fungus is often used against weeds. Work in this area has concentrated on fungal pathogens of plants as they can potentially be applied as sprays in the same way as conventional herbicides (hence their name myco- or bio-herbicides). Studies on bioherbicides have concentrated mainly on foliar treatments using fungi. Commercial products have been developed (mainly in the US) but success has been limited Soil micro-organisms are often overlooked but are also important as plant pathogens. Several are being investigated as potential biological control agents particularly for control of grass weeds such as downy brome, wild oat and green foxtail.

Although much of the work on biological control agents has concentrated on the growing weed plant, there is considerable potential for using micro-organisms to manipulate or deplete the soil weed seedbank. The persistence of weed seeds in the soil is the key to their success in continuing to emerge despite repeated control measures over many years. Greater predation or an increase in natural decay would reduce the soil seedbank and hence future weed populations. However, there are as of yet no practical or commercial applications available.

3. Conservation control is an indirect method, which manipulates the habitat around the weeds with the aim of encourging those organisms that attack the weed. This is a long term strategy that requires a detailed knowledge of the ecology of the crop weed habitat, the target weeds and the control agents. It has received little attention to date. One recent example is the upsurge of interest in looking at encouraging the dock beetle on dock plants by creating conditions that favour the beetle.

Livestock can also be considered as biological control agents which can give a broad spectrum control of weeds in various situations.

The assessment of the potential risks involved in introducing biological control agents remains a difficult and (sometimes) contentious issue as any predictions of how biological control may affect the interaction between species, and influence the life cycle of non-target species is extremely complicated. Even if there were no risk to non-target species, there could still be a conflict of interests because some may perceive a particular plant as a weed while others see it as a desirable wild flower, or even a potential crop. For this reason it seems difficult to imagine that of the shelf biological control of weeds is a realistic prospect in the short to medium term.

Allelopathy

Allelopathy can be regarded as a component of biological control in which plants are used to reduce the vigour and development of other plants. Allelopathy refers to the direct or indirect chemical effects of one plant on the germination, growth, or development of neighbouring plants. This can be through the release of allelochemicals while the plant is growing or from plant residues as it rots down. These chemicals can be released from around the germinating seed, in exudates from plant roots, from leachates in the aerial part of the plant and in volatile emissions from the growing plant. Both crops and weeds are capable of producing these compounds and in this case the desired effect is the impaired germination, reduced growth and poor development of weeds.

Potentially allelopathy could be used in various ways:

- To manipulate the crop-weed balance by increasing the toxicity of the crop plants to weeds thereby reducing weed germination in the direct area of the crop, which is the most difficult area to control physically.

- As cover crops to suppress weed germination and development over a whole field in part of a rotation.

- As mulched residues or incorporated residues which could prevent weed germination and allow transplanted crops to be grown, producing a residual weed control effect.

Many crops have been reported as showing allelopathic properties at one time or another and farmers report that some crops such as oats seem to clean fields of weeds better than others. The current list includes: wheat, barley, oats, cereal rye, brassicas, red clover, yellow sweet clover, trefoil, vetch, buckwheat, lucerne, rice, sorghum.

However, several weed species have also been reported to show allelopathic properties. They include couch grass, creeping thistle and chickweed. Where they occur together they may have a synergistic negative effect on crops.

Allelopathic effects might also depend on a number of other factors that might be important in any given situation:

- Varieties: there can be a great deal of difference in the strength of allelopathic effects between different crop varieties.

- Specificity: there is a significant degree of specificity in allelopathic effects. Thus, a crop which is strongly allelopathic against one weed may show little or no effect against another.

- Autotoxicity: allelopathic chemicals may not only suppress the growth of other plant species, they can also suppress the germination or growth of seeds and plants of the same species. Lucerne is particularly well known for this and has been well researched. The toxic effect of wheat straw on following wheat crops is also well known.

- Crop on crop effects: residues from allelopathic crops can hinder germination and growth of following crops as well as weeds. A sufficient gap must be left before the following crop is sown. Larger seeded crops are effected less and transplants are not affected.

- Environmental factors: several factors impact on the strength of the allelopathic effect. These include pests and disease and especially soil fertility. Low fertility increases the production of allelochemicals. After incorporation the alleopathic effect declines fastest in warm wet conditions and slowest in cold wet conditions.

Mulching

Mulches contribute to weed management in organic crops by reducing weed seed germination, blocking weed growth, and favoring the crop by conserving soil moisture and sometimes by moderating soil temperature. Opaque synthetic mulches like black plastic provide an effective barrier to most weeds and are amenable to mechanized application, but they must be removed at the end of the season. Organic mulches like straw suppress annual weed seedlings, conserve moisture, and add organic matter as they break down, but they are more labor-intensive to apply.

Mulching can reduce weed competition against vegetable crops, and save fuel and labor costs for weed control. Covering the soil surface with a suitable mulch can:

- Reduce weed seed germination.

- Shade and physically hinder emerging weeds.

- Enhance crop growth and competitiveness by conserving soil moisture and sometimes by modifying soil temperature.

Synthetic mulches like black polyethylene film (the most widely used plastic mulch) or landscape fabric are laid on a prepared seedbed just before transplanting or seeding a vegetable crop through holes or slits cut into the mulch. In-row drip irrigation lines under the mulch provide water and liquid fertilizers to the crop. Mechanization, with equipment such as tractor-drawn bed shapers, mulch layers, and planters, allows the farmer to mulch and plant a multi-acre field within a single day. Black plastic, other opaque materials, and infrared-transmitting (IRT) mulch effectively block weed emergence, and promote soil warming and early crop growth. Weeds emerging through planting holes may require manual removal, and alleys between mulched beds generally need cultivation or other weed control measures.

The National Organic Program (NOP) final rule (United States Department of Agriculture requires removal of plastic mulches from the field at the end of the growing season, and tractor-drawn mulch lifters are now commercially available to facilitate this chore. Despite the costs of capital equipment, the plastic itself, application, and removal, many organic vegetable farmers consider black plastic their most economical weed management option.

Figure: Black polyethylene film mulch gives these tomato transplants a head start by
blocking weeds, retaining moisture, and warming the soil

Organic mulches such as hay, straw, leaves, and chipped brush, are usually applied when the vegetable crop is well established and the soil has warmed to near-optimum temperatures. They are most effective on weeds emerging from seed, and least effective on aggressive perennial weeds emerging from rootstocks, rhizomes, or tubers. Organic mulch applied immediately after a final cultivation often suppresses later-emerging weeds until the crop has passed through its minimum weed-free period. Organic mulches generally lower soil temperatures and conserve soil moisture by slowing evaporation while allowing rainfall to penetrate. Normally, organic mulch is left in the field after harvest and, as it breaks down, it helps build soil organic matter.

Organic mulches. (a) About 3 inches of hay mulch have suppressed emergence from a large weed seed bank of galinsoga (Galinsoga spp.) and other annual broadleaf weeds in broccoli, onion, and garlic in the Appalachian region of Virginia. (b) Pepper thrives in a straw mulch in the Tidewater region of Virginia. At both sites, a few grasses and perennial weeds are beginning to break through, but the vegetables have benefited from soil moisture conservation as well as weed suppression by the mulch.

Manual application of hay and other organic mulches is labor intensive, and is practical only on a small scale. A few growers use bale choppers to mechanize application of hay or straw from small rectangular bales.

Many vegetable farmers apply straw or other organic mulches in alleys between plastic-mulched beds, either at planting or after cultivation. In addition to suppressing alley weeds, this system adds organic matter, helps conserve soil moisture and soil quality, and prevents excessive soil heating during summer, thereby realizing many of the benefits of both organic and synthetic mulches. The organic mulch can also improve fruit quality in pumpkin and other vine crops by preventing fruit-soil contact in alleys.

The organic mulch adds organic matter, conserves soil moisture, and prevents soil erosion in alleys. When hot summer weather arrives, the hay is pulled over the black film to prevent excessive soil heating, as shown here.

Figure: Alleys between plastic mulched beds are covered with a thick layer of hay to
suppress alley weeds in this pepper crop at Wheatland Vegetable Farms in Purcelle, VA

Mulches and Weed Seed Germination and Emergence

Light promotes seed germination in many agricultural weeds, including common lambsquarters (Chenopodium album), hairy galinsoga (Galinsoga ciliata), common chickweed (Stellaria media), common ragweed (Ambrosia artemesiifolia), common purslane (Portulaca oleracea), some pigweeds (Amaranthus spp.), black nightshade (Solanum nigrum), and annual bluegrass (Poa annua) (Mohler and DiTommaso, unpublished). Any opaque mulch, such as black plastic or several inches of hay, straw, or leaves, blocks the light stimulus, thereby reducing seed germination in these weeds after mulch application.

Seeds of an even wider range of common weeds respond to wide daily soil temperature fluctuations, including some that do not respond to light, such as horsenettle (Solanum carolinense), common cocklebur (Xanthium strumarium), and foxtails (Setaria spp.). Many summer annuals, including pigweeds, galinsoga, and purslane, germinate in response to high soil temperatures (85–100 °F). Organic mulches and white or reflective plastic films lower soil temperature and dampen daily fluctuations, thereby deterring weed seed germination.

Even with light and temperature stimuli blocked, a percentage of the weed seed population will germinate. However, the mulch intercepts light essential for photosynthesis and physically hinders seedling emergence. Dicot (broadleaf) seedlings are fairly delicate and easily suppressed by this mulch effect. Hay, straw, or cover crop residues at 3–5 tons per acre (2–4 inches, loosely packed) can prevent emergence of small-seeded broadleaf weed seedlings for at least several weeks, whereas a heavier mulch (7–10 tons per acre) may be required to block larger seeded species like common cocklebur or velvetleaf (Abutilon theophrasti), and some grasses, whose shoots are protected by a pointed sheath (coleoptile). Perennial weed shoots emerging from rootstocks, tubers, rhizomes, or bulbs can penetrate most organic mulches, and a few weeds, such as nutsedges, can puncture plastic film.

Dark colored synthetic mulches and IRT mulches increase soil temperatures and daily temperature fluctuations, which may stimulate weed germination. Since these mulches also block seedling growth, the net result is to draw down the weed seed bank.

Mulch Effects on Crop and Weed Growth

In addition to reducing weed seed germination and emergence, mulch can improve the growth and competitiveness of established crops by conserving soil moisture and modifying soil temperatures.

Soil warming under black or IRT plastic can enhance early season growth and maturation in heat-loving crops, while the soil cooling effect of organic and reflective film mulches benefits cool-weather vegetables like potato, and can help most crops thrive during hot summer weather.

Some organic mulches, such as hay, provide slow-release nutrients, or reduce certain pests by harboring their natural enemies. Reflective or colored synthetic mulches have been found to enhance the yields of certain crops by repelling pests or modifying the light environment around the crop.

It is important to note that, once a weed manages to emerge through the mulch, or emerges through a planting hole in plastic film, it enjoys the same soil moisture conservation and other mulch benefits as does the established crop. Conversely, any crop seedlings emerging beneath a mulch will be suppressed. Thus, it is common practice to spread straw or other organic mulches only after the crop is well established, and immediately after cultivation or manual removal of existing weeds.

Figure: Tomato grown in a non-irrigated field in the Tidewater region of Virginia thrived in hay mulch (right), and grew poorly without mulch (left). In addition to reducing weed emergence, the organic mulch conserved soil moisture and moderated soil temperatures

Mulching Limitations and Pitfalls

In some circumstances, mulching can aggravate weed problems. Organic mulches, especially hay from off-farm sources, may carry seeds of new weed species into the field. An organic mulch that is too thin to suppress weeds (e.g., 1–2 tons per acre, or an inch or so of material) may allow weed emergence, then enhance weed growth by conserving soil moisture. Legume residues have also been reported to release enough nitrate-N to trigger germination of nitrate-responsive weeds such as redroot pigweed (Amaranthus retroflexus). Aggressive perennial weeds can emerge through a heavy (6 inch) organic mulch, thrive, and steal moisture and nutrients intended for the crop. Weeds growing through mulch are more difficult to control mechanically, and may require special high-residue cultivators.

Untimely mulching, or using the wrong mulch for a particular crop, can slow its growth and leave it more vulnerable to weed competition. For example, fresh ("bright") grain straw can lower soil temperatures by as much as 10°F. Spreading straw or other soil-cooling mulches around newly-transplanted tomato or melon can set the crop back several weeks and give the weeds a head start.

Figure: Wild buckwheat (Polygonum convolvulus), a new weed in this garden, arrived in the mulch hay

Mulch materials with a high ratio of carbon to nitrogen (C:N ratio) have the potential to slow crop growth by immobilizing soil N. This is most likely to occur with finely divided materials (e.g., sawdust) or materials rich in soluble carbohydrates that can leach into the soil (e.g., sorghum–sudan greenchop). Coarse, dry materials like grain straw or chipped brush rarely tie up soil N unless they are incorporated into the soil.

Applying organic mulch around small, succulent lettuce, brassica, or other vegetable seedlings can result in defoliation by slugs or insects, leading to poor stands or delayed establishment. However, these organisms have been observed to attack weed seedlings as well, and can reduce weed populations without seriously impacting well-established crops. In addition, many organic mulch materials, especially freshly cut immature cereal grains, hay, or forage crops, may release substances that inhibit germination and seedling growth in both weeds and crops (allelopathy). Mulches applied when crops are well established will minimize these risks to crop production, and can provide mid- and late-season weed suppression.

Generally, organic mulches enhance moisture infiltration and reduce runoff. However, in situations where moisture is limiting, applying a thick, organic mulch on dry soil can prevent light rainfalls from reaching the soil and crop roots. In this situation, farmers can irrigate the soil thoroughly (to near field capacity), or install in-row drip irrigation lines prior to mulching.

Non-porous plastic mulches can hinder infiltration of rainfall or overhead irrigation into the crop root zone. Some water runs into planting holes, but much of it runs off the mulch into alleys and may not reach crop roots. Thus, almost all growers who use plastic install drip irrigation under the mulch to deliver water to the crop.

Other disadvantages of synthetic mulches include the labor of end-of-season removal, the petroleum embodied in the mulch, generation of non-biodegradable waste, and the fact that synthetic mulches do not add organic matter or nutrients to the soil. Compost or other solid organic fertilizers and amendments must be applied to crop rows prior to laying the mulch, and sidedressing the crop is limited to liquid fertilizers via the drip line and foliar feeding.

Even black plastic mulch will not give 100% weed control. Weeds, especially fast-growing viny species like morning glories (Ipomoea spp.) can emerge through crop planting holes, and require manual removal. A few aggressive perennial weeds like nutsedges (Cyperus spp.) can pierce synthetic mulches, compete with the crop, and complicate mulch removal.

Integrating Mulch with other Weed Management Practices

Mulching cannot alone provide sufficient weed control, and works most effectively in conjunction with other practices. For example, market gardeners often spread hay or straw after cultivating one or more times during crop establishment. Because organic mulches rarely block 100% of weed emergence, they give best results when used in conjunction with good crop rotation and measures to prevent or limit weed propagation. Similarly, measures to reduce populations of nutsedge, morning glory, and other aggressive weeds may be needed before synthetic mulches can be used successfully.

Figure: Weeds were successfully managed in carrot in this Floyd, VA garden through a combination of cultivation and mulching. Sown at the end of June, 2009, hoed twice during establishment, then mulched in early August (a), the crop remained mostly weed-free and closed canopy by early September (b)

Organic Mulching Materials for Weed Management

Organic mulches can suppress annual weeds and offer other important benefits, such as organic matter, nutrients, moisture conservation, soil protection, and moderation of soil temperature. Drawbacks include costs and labor of application, limited efficacy on perennial weeds, delayed soil warming, and the potential to carry weed seeds and harbor pests.

Hay, straw, and fresh-cut forage or cover crops are among the most versatile and widely-used organic mulches. They can suppress weed germination and emergence when applied at reasonable rates, are fairly easy to apply, reduce evaporative losses of soil moisture while allowing rainfall to reach the soil, and provide other benefits. Caution is needed to avoid bringing in weed seeds or herbicide residues with hay from off-farm sources. Tree leaves, chipped brush, and other forest-based mulches are often beneficial to small fruit and other perennial crops, but may not be an economical option for weed control at a multi-acre scale.

Organic mulch materials include grain straw, fresh or old hay, fresh-cut forage or cover crops, chipped brush, wood shavings, tree leaves, cotton gin waste, rice or buckwheat hulls, and other crop residues. Hay and straw are among the most widely used organic mulches in organic horticulture. Cover crops can be grown to maturity (flowering), mechanically killed, and left on the soil surface to provide in-situ organic mulch for no-till planting. Leaf mold (decomposed tree leaves), compost, and aged manure have also been used as organic mulches, although their crumbly texture may not provide as effective a barrier to weed seedlings as other materials.

Organic mulches suppress weeds in several ways. First, they block seed germination stimuli by intercepting light, reducing soil temperature, and greatly dampening day–night temperature fluctuations. As a result, fewer weed seeds germinate under the mulch than in uncovered soil. Second, the mulch physically hinders emergence of those weeds that do germinate. If the mulch is thick enough to prevent light from reaching the trapped seedlings, they eventually die. Third, some mulch materials, such as grain straw and fresh-cut forages like sorghum-sudangrass, release natural substances that inhibit weed seedling growth for several weeks after application, a process known as allelopathy. Finally, organic mulch can enhance crop growth and competitiveness against weeds by conserving soil moisture and moderating soil temperature.

Straw and other organic mulches effectively block emergence of most weeds germinating from seed, although grasses and large-seeded broadleaf weeds may require a greater thickness of material than small-seeded broadleaf weeds, which have more delicate seedlings. Perennial weeds arising from rootstocks, rhizomes, tubers, or other vegetative propagules can penetrate most organic mulches.

Weeds that have already emerged at the time of mulch application should be cultivated or hoed out before spreading mulch; simply laying the organic materials over established weeds is less effective. Once the weeds break through the mulch, they will enjoy the same mulching benefits as the crop, and will grow vigorously.

Usually, some weeds will eventually emerge through an organic mulch. Fast-growing, canopy-forming crop like sweet potato, squash, or snap bean often shade-out these late emerging weeds. In slower-growing, less competitive vegetables like onion and carrot, manual weeding or application of additional mulch may be required to maintain satisfactory weed control.

Hay

Hay is often used to mulch horticultural crops in regions such as southern Appalachia, where the predominant farming systems include hay production, and old hay is more affordable than straw and other materials. Hay has some drawbacks and must be chosen and used with care. However, it is fairly easy to apply in small scale plantings, and is usually beneficial to soil quality and crop production. A hay mulch of about 3–4 inch thickness can:

- Reduce emergence of weed seedlings, especially small–seeded broadleaf annuals.
- Provide habitat for beneficial organisms, including ground beetles and other weed seed consumers.
- Allow air and rain to reach the soil.
- Moderate soil temperature during hot weather.
- Conserve soil moisture.
- Prevent soil crusting and erosion.
- Keep pumpkins, melons, and other fruiting crops out of direct contact with soil, and therefore cleaner.
- Add significant amounts of organic matter and slow-release nutrients, especially potassium (K).

Figure: (a) Garlic mulched with hay immediately after planting in October; photographed in April. (b) Tomato mulched with hay after the crop became established, several weeks before the photo was taken. The mulch delayed weed emergence and provided favorable conditions for crop growth

Hay also has some significant drawbacks, in that it:

- Does not suppress most perennial weeds.

- May contain weed seeds or herbicide residues.

- Can harbor slugs, squash bugs, voles, and other pests.

- Can keep the soil too cool or wet, slowing crop growth or maturation.

- Can accentuate frost damage by keeping the soil's radiant heat from reaching crop foliage.

- Can build up excessive soil K levels when used year after year.

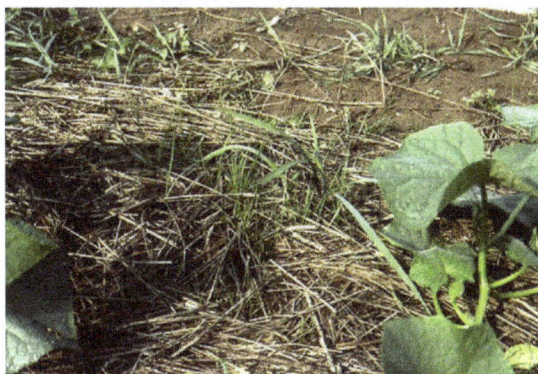

Figure: This hay was cut too late in its development, and carried mature seeds. As a result, forage grasses (fine–textured seedlings) are growing in this cucumber bed. Additional weeds have emerged from the soil's weed seed bank because the mulch application was not sufficiently heavy to cover the soil surface completely.

Sidebar: Hay Mulch

Hay from off-farm sources is a notorious source for new and serious weed species on a farm. Even in fields with good weed management, hay that has been cut too late in its development will carry seeds of the forage species themselves, which can be a nuisance if they come up in a vegetable crop. In addition to weed seed, the grower must be alert to the possibility of herbicide residues.

Some grass hay is produced with the use of weed control products that contain highly persistent active ingredients, including clopyralid, aminopyralid, picloram, and aminocyclopyrachlor, all of which are highly toxic to broadleaf plants. Hay from fields treated with any of these materials can cause severe damage to tomato family, cucurbit family, and other vegetable crops around which the hay is applied as mulch. Symptoms include curling and twisting of leaves and petioles (leaf stalks), and stunted growth, which can lead to crop failure or plant death. Subsequent vegetable or broadleaf cover crop plantings may continue to show symptoms for a year or more after initial contamination, and the field may lose eligibility for organic certification until herbicide residues have disappeared.

These herbicides are not degraded by composting. If horses or cattle graze or eat hay from treated fields, and their manure is hot-composted, cured for a year, and applied to vegetable beds, the vegetables can still suffer damage.

It always pays to check with the farmer who grew the hay regarding weed management practices, herbicide use, and time of cutting relative to forage seed set, before bringing hay onto the farm for use as mulch on horticultural crops.

Not all hay is alike. Grass hay is lower in nitrogen (N) and phosphorus (P), higher in K, and more persistent and weed-suppressive than legume hay. Because of its high cargon-to-nitrogen (C:N) ratio, grass hay has sometimes been reported to tie up soil N. However, this is most likely to occur when the hay is incorporated into the soil, not when it is applied to the surface as a mulch. A grass–legume mix (such as timothy–alfalfa, fescue–red clover, or rye–hairy vetch) yields a more balanced mulch that provides slow release nutrients to soil life and crops, and persists long enough to provide several weeks of weed suppression.

Fresh hay is more pleasant to spread but more likely to contain large numbers of viable weed seeds than old, spoiled hay. Second or third cuttings of hay are especially likely to have weed seeds (Mohler and DiTommaso, unpublished). Leaving hay bales or rolls in the rain for a year or so reduces weed seed viability, but moldy hay can be nasty and hazardous to handle, and does not provide as clean or long-lasting a mulch. A better solution is to grow and harvest mulch hay on farm, taking care to cut the mulch crop before viable seeds are formed. Mulch hay can be derived from perennial forages or annual cover crops (rye, sorghum-sudangrass, etc.). Note that repeated hay harvests from a given field can deplete soil nutrients, notably P, K, and calcium (Ca). Crop rotations that alternate annual or perennial mulch crops with vegetables that receive the mulch can promote better nutrient balance by minimizing the nutrient depletion of hay harvest while avoiding the potential K excesses from repeated mulch application.

Applying hay manually is most feasible at a small scale, for example, a half-acre of a high value crop. A few farmers have used bale choppers to mechanize application of hay or straw in small rectangular bales. Large rolls (round bales) are commonly unrolled between rows of widely spaced crops like tomato, a job which usually requires a tractor to place the ~1,000 lb roll at the beginning of the crop row, and two people to unroll it.

A number of farmers have streamlined on-farm harvest and application of mulch by using a flail chopper and forage wagon for harvest, and then pitchforking the fresh-cut forage off the wagon as it is pulled slowly along crop rows. Other producers, including David Stern of Rose Valley

Farm in upstate New York, grow alternate rows of vegetable and cover crop (e.g., potato and sorghum-sudangrass), and periodically mow the cover crop, blowing the clippings into the vegetable row as mulch. This approach saves the labor and costs of curing, baling, and storing hay. However, fresh grass or legume "green chop" has been reported to promote certain soil-borne pathogens for a short period after application (Mohler and DiTommaso, unpublished); thus, fresh-cut forage mulches should be tested on a small area for each crop before field-wide application.

Some tips for optimal use of hay mulch for weed control:

- Grow and use on-farm hay if practical.

- Check sources of off-farm hay for weed seeds and herbicide residues before purchasing.

- Apply mulch when crops are well established, and soil temperature and moisture are optimum for the crops being grown (exception: fall planted garlic is mulched immediately after planting).

- Hoe or cultivate at the beginning of a warm sunny day, wait 12–36 hours to let uprooted weeds die, then spread mulch (applies to all organic mulches).

- Use enough hay to suppress most weed seedlings, about 3–4 inches or 5–10 tons per acre.

- Monitor soil nutrient levels, especially K.

- Rotate mulched vegetables with non-mulched crops or hay production.

Straw

Straw is defined here as the stalks and other residues left after harvest of a mature grain, is similar to hay in texture, potential for soil protection and moisture conservation, weed suppression, and application methods. Straw differs from hay in that it:

- Has higher carbon to nitrogen (C:N ratio).

- Provides a cleaner, more persistent mulch that is slower to decompose, and more effective in keeping the fruit of pumpkin and other vine crops clean.

- May carry seeds of the grain crop itself, but is less likely to carry other weed seeds.

- Has somewhat lower K levels and slower K release.

- Is lighter colored and more reflective, hence it may cool soil more than hay.

Because straw is so much less likely to introduce serious new weed problems than hay, organic horticultural farmers located in or near grain-producing regions where straw is available and affordable often prefer straw over hay. The high C:N ratio of straw precludes much release of N from mulch to the current year's crop, but usually does not lead to tie-up of soil N, as long as the mulch lies on top of the soil and is not tilled in.

The dramatic soil cooling under straw can delay crop growth; however it can be beneficial for cool weather crops like potato, in which tuber growth is inhibited by soil temperatures above

70 °F and for other crops during hot summer weather. For example, tomato shows optimal nutrient uptake and production at root zone temperatures of 70–85 °F, and becomes stressed at higher temperatures; thus, it often performs better in organic than in plastic mulches during the heat of summer. Bright, reflective straw can intensify heating of crop foliage under a row cover, resulting in crop damage, and may also increase damage from frosts.

Figure: The light colored grain straw was applied too early in the season. The mulch has suppressed weeds, but also seriously delayed soil warming and tomato growth (compare to plastic mulched tomato in upper left)

In figure above (a) Potato tuber yields are often enhanced by the cooler soil conditions under a straw mulch. (b) The straw was applied after the soil had warmed to optimal temperatures for eggplant, and is now helping the crop thrive during intense summer heat. A few weeds have emerged at this point, but are unlikely to affect yield in the vigorous, established eggplant crop.

Rye, wheat, and other grain crops cut for mulch at an earlier stage of maturity (e.g., head emergence or pollen shed) are richer in nutrients and less likely to immobilize soil N than straw left by grain harvest. Rye, triticale, and other winter grains cut at the milk stage (before the seeds become viable) yields excellent straw for mulch, and minimize the risk of volunteer cereal grains becoming a weed. Cereal grain cover crops rolled down at the milk stage are particularly popular for no-till pumpkin production, as they help keep the fruit clean, reduce soil borne diseases, and promote even color development (Ron Morse, Virginia Tech, pers. comm).

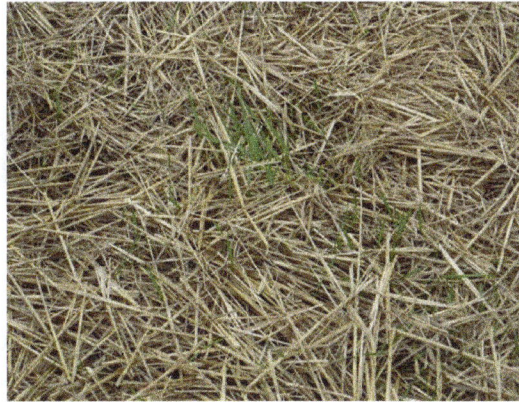

Self-seeding of cereal grain occasionally causes a weed problem in straw mulch.
In order to avoid this problem, some farmers grow their own grain straw for mulch,
and harvest the mulch crop before seeds become viable.

Tree Leaves

Hardwood leaves that fall naturally in autumn are sometimes used as mulch in vegetable production. They are rich in calcium (Ca) and micronutrients, contain small to moderate amounts of N, P, and K, and decompose gradually to form leaf mold, a humus-like material that is valued by horticulturists. Millions of suburban residents rake up autumn leaves for disposal, and a growing number of farmers and other entrepreneurs accept leaves for mulch or for making compost. Leaves are often used for berries and some other perennials that tolerate or prefer some acidity. Pine needles (pine straw) are lower in nutrients, more persistent, and more acidic than hardwood leaves, and can be especially useful for blueberries, which require a low pH. Tree leaves are much less likely than hay to carry the seeds of agricultural weeds; however, they have been observed to carry tree seeds (especially maple or ash), which germinate into vigorous seedlings that readily emerge through the mulch.

Figure: (a) Onion mulched with tree leaves gathered the preceding autumn at Potomac Vegetable Farms in Vienna, VA, near Washington, DC. (b) New blueberry planting in Floyd, VA mulched with pine needles (foreground) and grain straw (background)

Some disadvantages of tree leaves as a mulch include:

- A tendency to mat down when wet, creating soggy or airless soil conditions.

- A tendency to blow away in the wind when dry, or to blow onto and smother young crop seedlings.

- Labor intensive application, not feasible at a larger scale.

- Presence of trash (cans, glass, plastic, etc.) in municipal leaves or yard waste.

The soil benefits of tree leaves can also be realized by including them in compost piles, or making leaf mold (leaves aged for 1–2 years until crumbly), which is an excellent soil amendment or potting mix ingredient.

Chipped Brush, Wood Shavings and Bark

These forest product mulches are most often used on perennial crops such as berries and ornamental perennials, many of which like a somewhat acidic soil rich in mycorrhizae and other beneficial fungi supported by these materials. They tend to be coarser and higher density than hay or straw, require higher tonnage per acre to suppress weeds, and may not be economical for most larger-scale applications. Other characteristics include:

- High C:N ratio.

- Relatively long lived.

- Allelopathic properties when fresh, especially walnut and some conifers (softwoods).

- Provide calcium (Ca), micronutrients, and small amounts of N, P, and K.

- Formation of stable humus when fully decomposed.

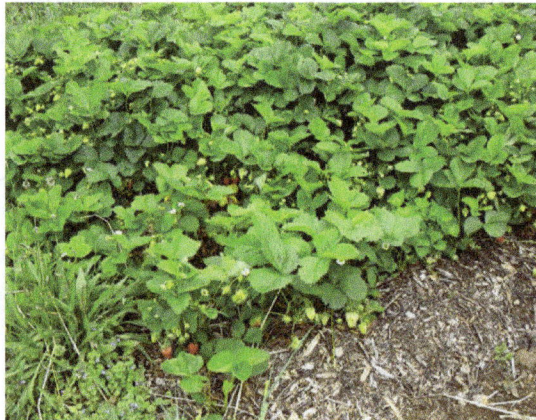

Figure. A perennial variety of strawberry in a garden in Floyd, VA thrives and yields well in a mulch of chipped brush, aged about one year before application

Wood based or bark mulches should be aged for at least a year outdoors before application near crop rows, to minimize possible allelopathic suppression of crop growth. However, fresh chipped brush can be useful for suppressing weeds in paths or alleys between beds. One grower in New Jersey has had excellent results with 1–2 year old hardwood chips as mulch, and 8–11 year aged hardwood chips as a soil amendment for blueberry.

Wood chip and bark mulches should not be piled against the bases of trees or shrubs, as this can promote the development of fungal diseases. Limit mulch depth to 1–2 inches adjacent to and within 6–12 inches from the base, then increase the depth further away.

Sawdust

Sawdust is chemically similar to other wood products, but because it is so finely divided, it has the following disadvantages as a mulching material:

- Tends to mat down and keep soil wet and airless.

- Can tie up soil N as small particles or soluble carbohydrates leach into the soil.

- Can be quite allelopathic against crops for a short time.

- May be penetrated by some weeds, and may provide a good germination medium for wind-borne weed seeds.

- May be washed away by heavy rain on sloping fields.

Figure: (a) An intense rainstorm has washed a fine sawdust mulch away from newly planted blueberries. (b) The same storm damaged soil structure in un-mulched beds (right and background), whereas chipped brush held firm, protecting both soil and crop (left foreground)

Compost

A few growers use compost as mulch, although the quantities required for effective weed suppression may not be economically feasible. In a study in Virginia, 1½–2 inches of leaf mold compost (50–90 tons per acre) did not suppress weeds quite as well as 4 inches (~8 tons per acre) of hay. Compost is much more effective and economical to use as an ingredient in potting mixes (at 10–50% of total volume), or as a soil amendment at 1–10 tons/ac to inoculate the soil with beneficial organisms, provide slow-release nutrients, and improve soil structure. Higher application rates, such as those used in the study, commonly leads to excessive levels of P, K, and some micronutrients in the soil. The surplus P and K can favor the growth of weeds over crops in subsequent years.

Figure: Mulching with compost (a) A municipal compost, based primarily on tree leaves, was applied at 50 tons per acre in this trial. (b) By midsummer, considerably more weeds emerged through the compost than through a 4–inch (~8 tons per acre) hay mulch

Manure

Manure is not recommended as a mulch for weed control. Many weed seeds pass through livestock digestive tracts unharmed, and the readily available nutrients in the manure stimulate weed growth. Lambsquarters (Chenopodium album) and spiny amaranth (Amaranthus spinosus) are just two of many nutrient-responsive weeds that are frequently spread in manure. Furthermore, uncomposted manure cannot be applied to USDA certified organic vegetable crops within 90–120 days of harvest, and applying sufficient manure to suppress weed seedling emergence from the soil is likely to create gross excesses of soil P and K.

Other Organic Residues

Crop residues—especially materials like cotton gin waste, rice hulls, peanut hulls, and buckwheat hulls—may be available in quantity in certain locales. Their ability to suppress weeds may vary, depending on texture and possibly chemical properties. Care should be taken to avoid crop residues that carry crop pathogens, weed seeds, or herbicide residues. Buckwheat hulls have been reported to attract cats using the mulched bed as a litter box, and thus may not be a good choice in neighborhoods with high cat populations.

Living Mulch

For many years, some growers have experimented with living mulches—perennial or annual cover crops growing between crop rows—in an effort to build soil quality while suppressing weeds. Experience has shown that living mulches allowed to grow in close proximity to crops often compete with the crop for moisture or nutrients, resulting in lower yields. However, in wide-spaced plantings, such as berries, alleys can be maintained in a perennial living mulch, while the area near crop rows are kept free of competing vegetation and mulched with straw, wood chip,or other organic materials. Living mulches can also be planted in 2–3 foot wide strips between permanent vegetable beds to create firm, mud-free paths for tractor and foot traffic; define where workers and u-pickers should walk, and provide habitat for beneficial insects. Clippings from periodic mowing of the living mulch can be used to supplement organic mulch in crop beds.

Figure: A perennial living grass–clover mulch

A perennial living grass–clover mulch. maintained by regular mowing, maintains soil quality and suppresses weeds in alleys, while a 4-ft-wide zone for each row of blueberries is kept free of competing vegetation and mulched with straw and clippings from the alleys, to allow the new planting to become established. The grass-covered alleys also provide a better surface for foot traffic and minimize soil damage in u-pick berry fields.

Cultural Weed Control

Organic farmers recognise that every element of farming is inter-linked, and that good rotational design produces healthy soil, healthy plants and good yields. Crop rotation is the cornerstone of organic farming practice. Rotation and forward planning are also important for managing weeds. In this topic we provide information on cultural methods aimed at preventing weed problems arising in the first place and which farmers and growers can plan to incorporate into their rotations.

The underlying principle of a preventative approach is to produce a constantly changing environment to which no single weed species can adapt and become dominant and unmanageable. In practical terms this means as diverse and long a rotation as possible consistent with the farm system and which prevents the weeds returning seeds to the soil seed bank.

Crop Rotation

Crop planning is a cornerstone of organic farming practice and it has important implications for weed management. It can be designed to positively influence weed control and to make a useful contribution to the whole farm management strategy. Typically rotation cycles extend over several years with often only an annual change of crop, but the inclusion of cover crops, intercrops and green manures increases the crop diversity in a rotation. In horticultural systems there may be sequential cropping where short-term crops follow each other in succession.

Weed population density may be markedly reduced using crop rotation but there has been little experimentation. Success depends on the use of crop sequences that create a diverse pattern of competition, allelopathic interference, soil disturbance, and production needs (such as the time of sowing and harvesting). There should be regular changes between spring and autumn-sown crops, and between annual and perennial crops, between dense leafy crops and those with an open habit, and between crops that require a long growing season and others that mature quickly. Rotation may also allow the use of a range of cultivations and direct non-chemical weeding methods that may be applicable to the different crops. The aim is to provide an unstable and inhospitable environment that prevents the proliferation of a particular weed species.

Choosing Crops and their Sequence

The length of the rotation, the choice and sequence of crops will depend upon individual farming circumstances that will include factors like soil type, rainfall, topography and enterprises. However, the aim is to produce an unstable environment in which no single weed species is allowed to adapt, become dominant, and therefore difficult to manage. No one rotation can be recommended, but ideally in terms of weed control rotations should include:

- Alternation of autumn and spring germinating crops,

- Alternation of annual and perennial crops (including grass),

- Alternation of closed, dense crops such as oats which shade out weeds, and open crops such As maize which encourage weeds,

- A variety of cultivations and cutting or topping operations that directly affect the weeds.

Various suggestions and observations include:

- Putting sensitive annual crops after perennial leys. Research has shown that in the third cropping year after a grass/clover ley there is twice as much weed emergence as compared to the first.

- Include a row crop that allows the use of one or more cultivations to kill emerged weeds and encourages the germination of others, so reducing the soil seedbank and hence potential weed numbers in future crops. Cultivations may also reduce the problem of perennial weeds by disrupting growth and smothering regeneration in the growing crop. Typical cleaning crops include turnip, sugar beet, and potato.

- Uncultivated leys provide a completely different habitat for weeds and may be used to reduce or eliminate particular weed species. Few studies have been made of the effectiveness of leys for controlling weeds but trials suggest that there is little advantage for weed management in leaving leys down longer than 3 years. The species composition, and the mowing and grazing regimes are important in the weed dynamics. Management of the weeds at the time of ley establishment is critical as is the method of ending the ley to avoid a flush of weeds due to the release of seed dormancy by cultivation. A greater proportion of ley in the rotation usually results in lower seed numbers in the seedbank in comparison with arable crops. It was a traditional way to deal with land infested with wild oat but does not eliminate the weed completely.

- Where a long grass break does not form part of the rotation weed problems are likely to be more severe. The problem will be greater where less vigorous and therefore less competitive crops are grown. Among the cereals, oats and winter rye are the most competitive followed by triticale.

- Canopy development and shading are important for weed suppression and choice of cultivar can influence this.

- Higher seeding rate and narrower row spacing increase the level of weed suppression.

- Competitive cereals like rye may be grown as short duration.

Fertility Building

The fertility-building period, or ley, will influence the weed population. If it is well managed it can act as a weed suppressing phase. It is important to choose the right species and ensure they establish quickly, especially for grassland systems where the ley may last for several years. Establishment of leys can be easier in the autumn period than in the spring because sowing in spring coincides with the main spring flush of weeds. The seedbed needs to be well prepared, and good contact made between seed mix and, ideally, moist soil to achieve good establishment. The choice of fertility building crop is also important. Rotations with grass leys have been shown to be beneficial in reducing weed seed numbers compared with rotations that do not include a grass phase. Grassland systems, which have temporary leys rather than permanent pasture, will provide the opportunity to control perennial weeds during the cultivations between ploughing and reseeding.

Grassland or clover/grass leys are an important part of the organic farming system in the UK. On livestock farms grassland forms the basis of the production process, in arable systems the ley is used primarily for maintaining or restoring soil fertility. The grass may be managed as a short, medium or long-term crop and this may determine the composition of the desirable sward species and the nature of the associated weeds. The seed mixture for a ley may include a relatively simple mixture of grasses and legumes or may be more complex and contain a range.

Choosing Varieties

Varieties that consistently suppress weeds are generally more desirable in organic systems (although this might be outweighed by marketing necessities) as opposed to varieties that tolerate weeds (and which potentially allow weeds to develop and return seeds to the soil seed bank).

Organic varieties: should show quick germination and establishment, rapid early and vigorous growth, and the ability to rapidly cover the soil and shade it (prostrate or tall varieties) to out-compete weeds at an early a stage in the crop cycle as possible. Varieties are well known to differ in architecture and competitive ability and whilst those that out-compete weeds are preferred it should also be borne in mind that those with erect foliage or that can tolerate some degree of mechanical weeding are also likely to be useful. Some crop types or varieties might produce allelochemicals that prevent weeds from developing or germinating although information on this is generally lacking.

In grassland systems the choice of variety may be dominated by forage value, but if there is opportunity the most vigorous species should be selected, as these will determine the productivity of the whole ley period. The trend in conventional cereal production has been to grow the taller stemmed varieties for their weed suppressing ability. Some farmers have stayed with the shorter stemmed varieties and employed a weed topper/cutter which will remove and, ideally, collect weed seedheads so long as there is a difference in height between crop and weed. New research in wheat is investigating leaf angle development, height and speed of development on weed suppression to aid farmer variety choice.

- Seed size: varieties with a larger seed size have been shown to exhibit greater initial vigour of emergence and growth, which may subsequently provide extra competitive ability. If there is a choice available then the most vigorous species should be selected, as these will be more likely to out-compete weeds and suppress their development.

- Clean seed: It is important that the crop seed is free from contamination by weed seeds. Organic farmers are obliged to use organically produced seed and this should be clean. It is important if saving seed that it is taken from weed free crops, and ideally, professionally cleaned. Tolerance levels of contamination should be low although they are not generally well defined as of yet. Tolerance levels of dock in Switzerland are one seed per 100g.

Seed Rate and Crop Spacing

Spatial distribution of the canopy foliage and rooting system will be important for weed suppression. In drilled or transplanted crops the proximity of the plants to one another will determine the competitiveness of the plant stand as a whole. The principle is that the greater the amount of space taken up by the crop in the rows, the less space there is available for the weeds to invade. However, it should be borne in mind that closely spaced crop plants compete with each other and that it is also expedient to allow sufficient space between plants to allow for efficient mechanical weeding should weeds develop and threaten the crop.

Seed rates tend to be higher for organic than conventional crops. There is also an allowance for potentially lower germination rates and loss of the crop by mechanical weeders.

There has been much work in cereals on row spacing, pattern, direction of sowing and seed rates (typically 10% higher in organic cereals). Results are varied and interactions between the factors are often significant. For example in narrow widths an E-W sowing was favourable whereas in wide row widths a N-S sowing showed better response. Findings from the EU funded WECOF trials are awaited to give more definite recommendations.

In vegetable crop market size specifications are often the main driver in determining the crop row spacing.

Establishing the Crop

The ability of the crop to get off to a good start ahead of the weed flora is critical. Good soil management practises are important to provide the best possible seedbed in which to plant a crop. The impact of a poor compacted soil can soon be seen on crop establishment and subsequent weed invasion.

In some systems sowing can be aided by the use of primed seed, or by transplanting an already established plant into a freshly prepared weed free seedbed.

Transplanting is a popular technique in organic horticultural systems. Bare rooted transplants can be raised on holdings or modular plants raised or bought in then planted out in the field.

Advantages include the benefit of accurate spacing i.e. not having to rely on germination that can sometimes lead to uneven establishment with subsequent yield and quality penalties. It also accentuates the difference in size between crop and weed, which can be vital for mechanical weeding at later stages.

Intercropping and Undersowing

Intercropping (or mixed cropping) and undersowing involves growing two or more different crop species in the same area. The advantage for weed control is that the crops cover more ground, so there is less space available for weed emergence. Intercropping can involve purely cash crops or a mixture of cash crops and fertility building crops.

- Intercropping: Is widely practised in certain countries and an enormous variety of intercropping systems have been developed. Both component crops may be taken to yield or one may be there as a living mulch to improve weed control. In successful intercrops weed suppression is usually superior to that of either of the component crops when grown alone. Crop density, crop diversity, crop spatial arrangement, choice of crop species and cultivar will all affect weed growth in intercropping systems. If water or nutrients are limiting then growth of one or both intercrops will suffer. Improved weed control alone is unlikely to justify their use and there must be other obvious benefits if the change in cropping practice is to prove economic.

In terms of mixed cash cropping there have been investigations into organic winter wheat and beans that reduced weed growth and gave better yields than sole cropping. Leeks and celery intercropping has also been shown to increase weed suppressing ability, and reduce reproductive capacity of late emerging groundsel. There are probably a wide range of combinations that could be designed to suit the farmer's rotations and marketing needs although some practical experimentation is required

- Undersowing: Aims to cover the ground with a quick-growing dense layer of vegetation underneath the crop. The undersown species is prostrate, usually leguminous, and adds to or maintains fertility. It also suppresses weeds. Combining cash cropping with fertility building in this manner potentially produces an economic return, and it may mean there is either no need for an isolated fertility period, or that the length of that phase can be reduced. Undersowing cash crops with fertility building crops has other advantages apart from weed suppression, but so far the technique has only been widely used in cereal growing where it helps to re-establishing leys and avoid bare ground after harvest. Short-strawed cereal varieties can be difficult to manage, and straw difficult to save, due to the undersown crop growing up into the cereal. Further research into variety choice and sowing rates is needed to ensure that competition between the two crops is kept to a minimum.

Using Cutting Regimes for Weed Control

Cutting and topping weeds will have an impact on the type and weed flora in a field and can be invaluable in preventing return of weed seed to the soil seed bank. Cutting and topping are important for weed management in pasture, grass and leys. Topping can also be used as a remedial measure in vegetable or other crops to prevent weeds from seeding.

- Pasture systems: Good management involves maintaining the condition of the sward by cultural means. In particular reducing weed intrusions by chain harrowing in spring and topping regularly during the growing season. Mowing during the seeding year must be carefully judged and close cutting avoided. Spring sown stands should be cut no later than mid-August to allow recovery before winter. Summer sowings should be left unmown until November. Undersown lucerne should be left to grow into the winter. Where companion grasses are growing strongly, light winter grazing may be desirable. In grazed pasture weeds that are not eaten by livestock, will need to be topped to prevent seed shed. The established crop may be cut up to four times per year starting in mid-May. The crop is quickly weakened by defoliation, either by grazing or cutting, at too young a stage especially in spring or autumn. Before entering the dormant stage the crop must be allowed to make sufficient growth to replenish the food reserves in the root.

- Grassland systems: Cutting for hay or silage will have an impact on the weed flora. Silage tends to be cut early in the season when the sward is young and fresh, whilst hay is cut at a later stage. There can be both advantages and disadvantages associated with the timing of cutting depending on the weed flora and the ultimate requirements of the system. Cutting late may allow weeds in the pasture to grow to maturity and set seed. The ripe seeds may contaminate the hay and remain viable when passed through livestock. Dock seeds should not survive low pH silage, however they will survive in a later cut of hay. This mature seed may also shed on the ley surface and find opportunities to germinate in situ or be transported by livestock to other locations. In contrast, cutting early for silage in fields, with for example an infestation of creeping thistle, may encourage the spread and growth of this weed. Hence, there has to be a balance between the requirements of the farming system and weed control implications.

- Horticultural and stockless arable systems: Ley management will include topping at intervals during the summer to a height of around 10-15 cm. Ideally in fertility building leys the sward should not be allowed to get higher than 40 cm (or knee height). If the vegetation gets higher than this, then topping will create a mat of vegetation that will act like a mulch. This can create dead spots in the ley where clover may be excluded by the more vigorous grasses, or which weeds may colonise. Topping the ley regularly will also ensure that tall weeds that may have germinated will not be able set seed.

Using Manures

The use of raw manures and slurry has often been associated by farmers with increased weed problems. The problems can arise in various ways; either as a result of weed seeds in the manure, as a result of the way in which it is applied or due to the stimulatory effect of the nutritents on weeds already present in the soil.

- Weed seeds in manure: Some manure contains weed seed, either seed that has passed undigested through animals or from bedding materials like small-grain straw and old hay. High-temperature aerobic composting (recommended under organic standards) can greatly reduce the number of viable weed seeds as long as the temperature is maintained at higher than 60°C for more than three days. Operationally compost will need to be regularly turned to achieve even heating through the whole heap and to get material from the outside (where seeds are likely to survive) to the inside (where the highest temperatures are likely to be generated. In a similar way, aeration of slurry can reduce the number and viability of weed seeds.

- Applying manure: When applying manure or slurry try not to create conditions which stimulate weed seeds to germinate (excessive soil disturbance, creating bare patches etc.). Applying slurry to stubble after silage cuts can provide optimum conditions for weed seed germination. A nutrient-rich bed of cattle slurry will produce a high potassium environment which will favour weeds such as docks rather than grasses. Dock seeds should not survive low pH silage but will survive in a later cut of hay.

Some research has shown that placing manures and slurry more accurately on crops can benefit the crop rather than the weeds. Crop plants are generally sown fairly deeply and they germinate from a lower level in the soil profile than weeds, which tend to dwell on the surface and germinate from 0-3 cm. Crop plants also root more deeply. This tendency can be exploited for weed management. In arable/horticultural systems manure placed 10 cm below the soil surface encourages the crop seeds to grow down into the nutrient-rich layer before the surface-dwelling weeds can reach it. This technique can also be used with broadcast spread slurry. If it is ploughed in rather than left on the surface it will be available to the crop before the weeds can reach it.

In many cases, the growth of weeds that follows manuring is a result of the stimulating effect manure has on weed seeds already present in the soil. This can be due to the flush of nutrients (e.g. supply of nitrates), enhanced biological activity in the soil, or other changes in the fertility status of the soil. Some work has indicated that excesses of potash and nitrogen in particular can encourage weeds but in any case it is prudent to monitor the nutrient content of your soil and manure, and spread manure evenly to reduce the incidence of weed problems.

Don't give the docks an advantage: applying slurry to stubble after silage cuts can provide optimum conditions for weed seed germination. A nutrient-rich bed of cattle slurry will produce a high potassium environment which will favour weeds such as docks rather than grasses. Dock seeds should not survive low pH silage but will survive in a later cut of hay.

Weed Management and Livestock

In mixed systems, where grass/clover leys are used for fertility building, livestock can make good use of the nutrients and they also produce manure, a resource which can be used around the farm to fertilise cash crops. Apart from leys, pastures will also need, weed management and increasingly conservation of old or rough pastures requires specialist grazing. Animals can also be used to consume cut weeds or other plant material like chaff or screenings that are likely to contain some weed seeds.

Animals have different grazing habits and it is even recognised that different breeds or individuals are likely to have different tastes and habits. The species, breed, age and individuality of animals will all affect what they will eat and therefore what effect they will have on both weeds and pasture. Variability within the feeding site (e.g. vegetation, topography) can also be important as can other factors such as the weather. In general terms:

- Goats are browsers and have a reputation for enjoying tough and woody plants.

- Sheep are recognised as being useful for weed control as they graze close to the ground and will eat a wide range of plants. They can be used early and some breeds are hardy.

- Cattle can be used for early grazing but there are a large number of different breeds and types with different grazing requirements including beef, dairy and traditional breeds. Grazing strategies appear to be related to plant energy content and digestibility and this will affect how plants are eaten (leaves or stems or other parts of plants), which plants are eaten (species) and size plants eaten (young or more mature plants). Cattle tend to avoid longer coarser grass and hairy, spiny or poisonous plants. The selection of certain plant species and plant components as well as the location of these plants is based on the previous experience of the animals or learned from their mothers when they are calves.

- Pigs are good at rooting and have been recommended at various times for digging out perennial weeds like dock and couch when fenced within fields (and tightly stocked).

- Geese consume grassy weeds and have been used to weed in between rows of well -established crops.

- Horses and ponies are grazed on ever increasing areas of land and can be used as part of a grazing rotation. They prefer frequent small amounts of fibrous grass or other high roughage material. They have been known to dramatically increase the number of docks in a rotation.

It is important to get the right grazing balance over the year to get the maximum benefit for the animals and also to prevent damage to the sward or soil. For example, stocking more lightly in the winter months and in wet periods prevents poaching. So think about the right season to graze, how long to graze, how many animals to graze and how long the grazing area will need to recover. Things to consider include:

- Timing grazing to benefit the pasture and promote competition with weeds.

- Timing grazing to damage the weeds, e.g. to remove flowers or seed heads before seed production.

- Allowing time for the pasture or forage to recover between grazings.

- Making sure that livestock that have been grazing on weedy land feed on weed seed free forage for 4- 5 days before introducing them to weed free areas or pastures (some seeds will remain viable after passing through animals which may take a few days).

Suggestions for rotating livestock, depending on situation, include:

Alternate the grazing of sheep and cattle from year to year or to use mixed grazing for better weed control. Mixed grazing in the same field may be detrimental to the cattle exploit animals.

Fallowing

It is usually not desirable to have to plan a fallow period into a rotation, but it may be necessary if weeds cannot be controlled during cropping or fertility building. It may not be necessary to stop cropping for a whole year, but instead to employ a bastard fallow i.e. no crop for part of the year. Tillage without a crop for a season is sometimes referred to as a black fallow. Fallowing the land for part of the growing season, as a bastard or summer fallow, can be as effective as a full fallow, is more suitable for lighter land and can be fitted into most rotations.

Fallowing is often best during the summer when cultivations can take place and the drier periods allow for root desiccation. This technique is more useful in plough-dominated systems rather than grassland management. One aim is to cultivate the soil progressively deeper over time, exposing underground plant parts to desiccation at the soil surface but in this case dry weather conditions are essential. Ploughing begins in June/July allowing time for an early crop to precede it. A bastard fallow is often used after a ley to reduce perennial weeds before sowing a winter cereal. There is also an opportunity for birds to feed on wireworms exposed during soil disturbance.

Fallowing has been shown to reduce perennial weeds within a rotation. The aim is to kill the vegetative organs of the weeds by mechanical damage and desiccation. For a full or bare fallow, heavy land is ploughed in April to give the weeds time to start into growth. It is cultivated or cross-ploughed 10-14 days later to produce a cloddy tilth. The soil is cultivated or ploughed at frequent intervals to move the clods around and dry them out. By August the clods should have broken down and the soil is left to allow the weed seeds to germinate. In September/October the weeds are ploughed in and the land prepared for autumn cropping. If a cereal is to follow the fallow, wheat bulb fly may be a problem because it lays eggs on bare ground in July. This can be overcome by sowing a green manure such as mustard to cover the land during this period.

Although there is the benefit of reduced weed control costs in subsequent crops after an effective fallow, the economics of taking land out of production for a full year together with undesirable effects on the soil and the environment, make the use of a bare fallow unlikely for weed control in the organic system. There is no financial return during the fallow period while labour costs accumulate during the fallowing operations. As an alternative to fallowing, cleaning crops such as potatoes and turnips allow repeated hoeing for weed control (but are not suited to heavy land).

A similar effect to that of fallowing can be achieved with rapidly developing crops like radish (Raphanus sativus) that are harvested before the onset of weed competition. The short interval between crop establishment and harvesting in this crop encourages weed seed germination but does not allow the weeds time to set seed or reproduce vegetatively.

Farm Hygiene

Weeds are, by their nature tenacious and almost impossible to eradicate once established. The best form of management is preventing their establishment in the first place. Weeds are easily spread between fields and between farms and it is worth taking some trouble to try and prevent this with some basic hygiene measures.

Have you got a system for detecting weeds early:

- Managing a particular weed will be easier if it is detected early and prevented from spreading.

- Ensure that all people who work on the farm or visit it are alert to the possibility of spreading weeds and weed seed and ask them to tell you if they notice any particular areas of weeds.

- Keep records of problems weeds and their spread or otherwise. A digital camera can be a useful tool to record presence of weeds and monitor changes over time.

The farm machinery spreading weeds:

- Weed seeds are easily carried in soil, crop residues and on machinery so these should be regularly cleaned down.

- If there is a serious weed infestation in a particular field, or machinery is moving through fields where weeds are flowering, then washing down machinery should be a serious consideration.

- Hygiene is particularly important at harvest time. In crops like cereals weed seeds may be scattered in the field or caught on the machinery and dislodged later some distance from the original source. It may be necessary to add screens to combines to catch weed seeds at harvest. Older models may already have these features.

Organic No-till

Since the advent of no-till in conventional row crop production, soil conservation and improvement aspects of no-tillage systems have attracted the interest of some organic farmers. The big question, of course, is how to do it without synthetic herbicides. The first big breakthrough occurred in the 1980s with the discovery that certain winter annual cover crops, notably cereal rye and hairy vetch, can be killed by mowing at a sufficiently late stage in their development—full head emergence with pollen shed in cereal grains, and full bloom in legumes. When winter annual cover crops at this stage of development are cut close to the ground, they generally do not regrow significantly, and the clippings form an in situ mulch through which vegetables can be transplanted with no or minimal tillage. The mulch hinders weed seed germination and seedling emergence, often for several weeks. This strategy is sometimes called "organic no-till," although continuous no-till is usually not feasible at this time for organic production of annual crops.

Figure: Summer squash (left) and broccoli (right) were planted no-till into a winter rye–hairy vetch mulch

In some initial experiments, the mulch effect of the mechanically killed cover crop was sufficient to delay the onset of weed growth until after the crop's minimum weed-free period, which made post plant cultivation, herbicides, or hand weeding unnecessary. Crop yields were commensurate with yields in control treatments in which the cover crop was incorporated. Tomato and some late-spring brassica plantings did especially well, and some large-seeded crops can be successfully direct-shown into cover crop residues. Such results, combined with research findings that no-tillage systems with cover crops can substantially rebuild soil organic matter and soil quality, stimulated widespread interest in developing organic no-tillage systems.

The cover crop generated over three tons aboveground dry weight per acre, and was either mechanically rolled (left) or manually cut with a sickle (right) and left in place, providing sufficient mulch to suppress weeds throughout the crops' minimum weed-free periods. No post plant weeding, cultivation, or herbicide applications were done in these crops.

Challenges in Organic No-till

Several problems dampened the initial enthusiasm for organic no-till. First, results were inconsistent, and both weed control and vegetable yields sometimes fell short of the standard set by vegetables planted after cover crops were tilled in as green manures. Weed suppression failed particularly when cover crop biomass was insufficient to provide a thick mulch, or when many perennial weeds were present.

Figure: Canada thistle, an invasive perennial weed, easily emerged through a cover crop mulch to compete with this no-till planted broccoli. The farmer has since brought this infestation under control through a combination of timely tillage and high biomass cover crops

In addition, planting vegetables through mulch can delay vegetable growth and maturity, or promote problems with slugs, cut worms, or certain crop diseases. Lower soil temperature under the mulch slows mineralization of soil nitrogen (N), which can reduce yields of some crops, especially broccoli and cool season greens, which need a lot of N over a relatively short period of time. Furthermore, the randomly-oriented cover crop residues left by most mowers, scythes, and other manual cutting tools, interfere with mechanical no-till transplanting, and most standard vegetable planters and transplanters do not function well in untilled soil.

Figure: These winter cover crops were cut manually with a sickle prior to manual planting of the broccoli crop. Mechanical transplanters, even those designed for no-till applications cannot operate effectively in randomly oriented residues like these

Finally, even when the cover crop residues eliminate early-season weed competition without hindering crop yields, later-season weeds almost invariably emerge, making tillage necessary after vegetable harvest to prepare a seedbed for the following cover crop. Thus continuous organic no-till is generally not practical, and some tillage is usually needed at least once per calendar year to manage weeds in organic annual cropping systems.

Advances in Organic No-till

In recent years, technical advances, combined with additional success stories from some working organic farms, have stimulated continued experimentation by farmers and researchers across the United States. No-till transplanters, seeders, and planting aids (coulter followed by a shank to prepare a narrow, deep slot of loosened soil for planting vegetables) have been developed that permit mechanized applications. Whereas the randomly oriented residues of sicklebar- or rotary-mowed cover crops may cause clogging problems with these implements, other options such as flail mowing, rolling, and roll-crimping show promise. The flail mower chops the residue fine enough for the no-till planters to penetrate the mulch to plant seeds or starts into the soil, though the finely chopped residues break down faster so that their weed suppression effect is shorter-lived. Rolled residues are oriented parallel to the direction of travel, thereby minimizing interference with mechanized planting.

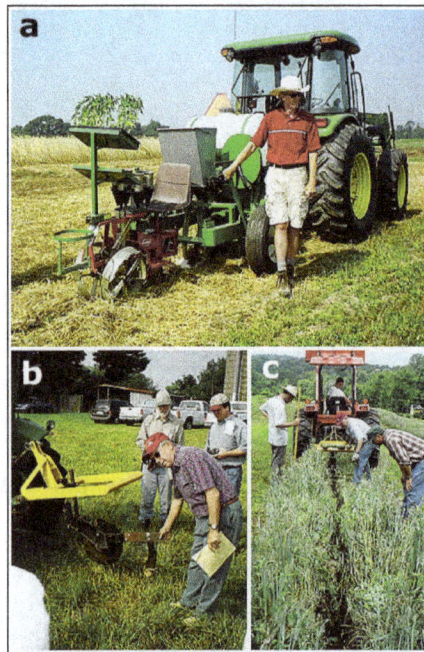

In figure above (a) Dr. Keith Baldwin of North Carolina Agriculture and Technology State University demonstrates a no-till vegetable transplanter, developed By Dr. Ron Morse at Virginia Tech and shown here in operation in a cover crop mulch at NCA&TSU's field station in Greensboro, NC. The planter parts residues, loosens soil in a narrow slot to 6–8 inches, plants and waters seedlings, and firms soil around seedlings. It is a major capital investment (at least $7000 per row). (b) Dr. Morse demonstrates a simple, inexpensive, light-duty no-till planting aid consisting of a coulter and shank to part residues and prepare a slot into which seedlings or large seed can be set manually or with standard planting equipment. (c) A no-till planting aid operating through still-standing cover crop prior to planting potatoes or large seeded vegetables. The cover crop is then mown just before the vegetable emerges to suppress weeds for maximum duration during the vegetable's growth.

In figure above (a) Cover crop residues that have either been chopped into fairly small pieces by a flail mower, or rolled so that their stems are oriented parallel to the direction of travel, permit mechanized no-till vegetable planting or transplanting. The flail mower shown chopping the cover crop here can also be drawn through the field with the PTO turned off to roll the cover crop (left foreground); thus, it is a versatile tool for no-till cover crop management. (b) Flail mower (PTO off) rolling a dense stand of winter rye–hairy vetch to form a thick mulch with residues oriented to allow mechanized planting.

Researchers have also found that winter cover crops are not the only ones amenable to no-till, no-herbicide management. Oats, field peas, and some vetches planted in early spring can be killed by mowing or rolling in early summer; some warm-season cover crops like soybean, foxtail millet, pearl millet, and sunnhemp can be similarly terminated about two months after seeding. The soil cooling effects of these mulches for midsummer and early fall vegetable plantings can be advantageous during hot seasons.

Non-winter-hardy cover crops planted 60–90 days prior to anticipated frost-kill can form an in situ mulch that suppresses winter weeds and lasts into the following spring. While attempts at no-till spring vegetable planting into winterkilled cover crops have been hampered by slow soil warming and inadequate weed suppression later in the spring, reduced tillage methods (strip till, ridge till, or shallow tillage) have given more promising results, with good vegetable yields and less problems with weeds compared to the no-till option.

Figure: This sudangrass cover crop, planted in July on Cape Cod, MA, winter-killed to form an in situ mulch that suppressed winter weed growth.

Although continuous organic no-till does not yet appear feasible, significant opportunities exist to reduce tillage in organic production, thereby conserving soil organic matter and soil quality, and possibly improving weed management. Within this context, no-till cover crop management and vegetable planting can be effective tactics when used at certain stages of a crop rotation.

By minimizing soil disturbance and exposure of weed seeds to light, rolling or mowing a cover crop rather than tilling it in helps to close the niche for annual weeds between a cover crop and the subsequent vegetable. Furthermore, surface residues of allelopathic cover crops like rye can maintain a shallow, weed-suppressive zone in the top inch or so of soil for some weeks. Larger seeded crops like beans, peas, and sweet corn, and especially transplanted vegetables, are quite tolerant to the allelopathic effects; hence the cover crop mulch acts to some degree as a selective herbicide. Transplanted tomatoes are especially tolerant to rye allelopathy, and respond positively to substances released by hairy vetch residues. In several studies, tomatoes grown no-till in rye–vetch have out-yielded tomatoes in other systems, including plasticulture with drip irrigation.

For small-seeded vegetable crops that are sensitive to allelopathy and require a fine seedbed, cover crops can be managed with ridge tillage, or strip tillage with sweeps ahead of the tillage implement to clear residues from crop rows. This provides a clean seedbed within crop rows and leaves the mulch between rows to suppress weeds. Ridge or strip tillage also allows prompt soil warming within crop rows, which is important when the farmer plants warm season vegetables early in order to capture early, lucrative markets.

Figure: Strip tillage prepares a narrow (~8 inch) swath through a cover crop mulch, allowing soil warming within the crop row while maintaining weed-suppressive mulch between crop rows

Considerations in Deciding Whether to attempt Organic No-till

No-till cover crop management and vegetable planting are most likely to work well in fields where:

- The cover crop is mature (heading/flowering with pollen shed), nearly weed-free, and has developed at least three tons (dry weight) aboveground biomass (solid stand three to four feet tall; cannot see the ground when viewed from above; clippings from one square yard weight about one and a half pounds when thoroughly dried).

- The cover crop includes a cereal grain or other grass that forms persistent mulch.

- Nutrient release from mulch decomposition approximately parallels vegetable crop nutrient needs, or sufficient supplemental N and other nutrients are provided.

- Weed populations are light to moderate, with predominantly annual broadleaf weeds.

- The soil is of light to medium texture, well-drained, and quick to warm up in spring.

- Moisture levels are adequate but not excessive, or can be supplemented by in-row drip irrigation.

- Organic mulches have historically been found to harbor natural enemies of important vegetable pests.

- Transplanted or large-seeded vegetables will be planted.

- Early vegetable maturity is not required.

- The farm has access to suitable equipment for rolling/mowing the cover crop and no-till planting through heavy residues.

No-till cover crop management and vegetable planting are not recommended when:

- The cover crop is not mature, is weedy (more than 5–10% of the aboveground biomass is weeds), or has not developed sufficient biomass (looks thin, can see patches of bare ground when viewed from above, dry clippings from one square yard weigh less than one pound).

- The cover crop is likely to decompose too rapidly to provide weed suppression (buckwheat or all-legume cover crops usually break down rapidly).

- Slower N mineralization in the mulched, untilled soil is likely to cause N deficiency in the vegetable crop.

- Invasive perennial weeds are present, the soil's weed seed bank is large, or annual weed poplulations are high and dominated by grasses or large seeded species that readily penetrate mulch.

- The field has been converted from sod to annual production within the past 12 months (bits of sod can regenerate and become perennial weeds under no-till without herbicides).

- The soil is heavy or clayey, and tends to be slow-draining or slow to warm up, especially in wetter-than-normal years.

- Slugs, squash bugs, or other pests that are commonly associated with organic mulch have historically been a problem.

- Small-seeded vegetables will be direct-sown (sensitive to allelopathy, slugs, and other mulch effects).

- Quick maturation of tomatoes or other vegetables is desired for an early market.

- The farm does not have the equipment needed for planting vegetables through a rolled or mowed cover crop at the scale of operation.

Compared to green manuring (soil incorporation of cover crops), no-till cover crop management slows the rate of N mineralization by keeping the soil cooler, and by leaving the organic residues on the soil surface. This can lead to N deficiency in a cool-season, heavy-feeding crop like spinach or broccoli that requires a lot of N in a short time, especially in heavier soils. On the other hand, the slower N mineralization under a no-till managed cover crop can be advantageous for summer vegetable production on sandy soils in warm climates. In one trial, summer squash yield doubled in mow-killed rye–vetch compared to squash grown after the cover crop was tilled in. The farmer suspects that the no-till system released N roughly as the crop required it, whereas much of the N released from the tilled cover crop leached away before the squash could utilize it.

Organic Herbicides for Weed Control

Organic herbicides kill weeds that have emerged but have no residual activity on those emerging subsequently. Further, while these herbicides can burn back the tops of perennial weeds, perennial weeds recover quickly.

These organic products are effective in controlling weeds when the weeds are small but are less effective on older plants. In a recent study, we found that weeds in the cotyledon or first true leaf stage were much easier to control than older weeds. The control ranged from better than 60% to 100% if these weeds received high volumes of these materials when they were just 12 days old. When broadleaf weeds were 26 days old, even high volumes of these materials gave at best less than 40% control.

Table: Broadleaf (pigweed and black nightshade) weed control (% control at 15 days after treatment) when treated 12, 19 or 26 days after emergence.

	Weed age		
	12 days old	19 days old	26 days old
Green match Ex 15%	89	11	0
Green match 15%	83	96	17
Matran 15%	88	28	0
Acetic acid 20%	61	11	17
Weed zap 10%	100	33	38
Untreated	0	0	0

Table: Grass (barnyardgrass and crabgrass) weed control (%control at 15 days after treatment) when treated 12, 19 or 26 days after emergence.

	Weed age		
	12 days old	19 days old	26 days old
Green match Ex 15%	25	19	8
Green match 15%	42	42	0
Matran 15%	25	170	0
Acetic acid 20%	25	0	0
Weed zap 10%	0	11	0
Untreated	0	0	0

We also found that broadleaf weeds were easier to control than grassy weeds — the best control on even young, 12-day-old grass weeds was only around 40 percent. This may possibly be due to the location of the growing point (at or below the soil surface for grasses) or the orientation of the leaves (horizontal for most broadleaf weeds).

All of these materials are contact-type herbicides and will damage any green vegetation they contact. However, they are safe as directed sprays against woody stems and trunks. For turfgrass sod production, organic herbicides could be applied when preparing the seedbed and then again with the first flush of weeds. Grass seed could be planted a bit deeper (1/4 to 1/2 inch deeper) to delay turfgrass emergence, so that the organic herbicide could control the broadleaf flush without adversely affecting the turfgrass.

Application

Organic herbicides kill only contacted tissue so good spray coverage is essential. For example, a large, flat nozzle would be preferable in turfgrass production. In tests comparing various spray volumes and product concentrations, high concentrations at low spray volumes (20% concentration in 35 gallons per acre) were less effective than lower concentrations at high spray volumes (10% concentration in 70 gallons per acre). Because organic herbicides lack residual activity, repeat applications will be needed to control new flushes of weeds.

In addition to high volume, we found that adding an organically acceptable adjuvant resulted in improved control. Among the organic adjuvants tested thus far, Natural wet, Nu Film P, Nu Film 17 and Silwet ECO spreader have performed well. Although the recommended rate of these adjuvants is 0.25 % volume per volume (v/v), increasing the adjuvant concentration up to 1% v/v often leads to improved weed control, possibly due to better coverage. Work continues in this area, as manufacturers continue to develop more organic adjuvants.

Environmental Conditions

Optimum environmental conditions are required when applying these organic products for good control of weeds. Temperature and sunlight have both been suggested as factors affecting organic herbicide efficacy.

In several field studies, we observed that organic herbicides work better when temperatures are above 75 °F, so applications in the winter may be less effective than summer applications. However, recent experiments have assessed winter weed control during cool conditions, and in spite of cold temperatures, plantain control was very good with Weed Pharm, or the high rates of Weed Zap or Biolink. Annual bluegrass control was also good with these same materials during cool conditions.

Table: Plantain and annual bluegrass control (%) at 4 and 9 days after treatment (DAT). Application made on January 6, 2011 during cool conditions (40 °F). All treatments included Eco silwet 0.5% v/v.

Treatment	Plantain control		Annual bluegrass control	
	4 DAT	9DAT	4DAT	9DAT
Biolink 3% v/v	52	48	15	35

Biolink 6% v/v	63	80	40	63
MOI-005 5% v/v	2	13	0	2
MOI-005 10% v/v	10	20	0	3
Green Match 7.5% v/v	12	13	3	5
Green Match 15% v/v	23	38	10	52
Matran 7.5 % v/v	5	8	2	3
Matran 15% v/v	20	17	5	30
Weed zep 7.5% v/v	18	28	10	42
Weed zep 15% v/v	52	78	23	78
Weed pharm 100%	82	90	53	87
Untreated	2	0	0	0
LSD.05	23	19	13	29

Sunlight has also been suggested as an important factor, and anecdotal reports indicate that control is better in full sunlight. However, in a greenhouse test using shade cloth to block 70% of the light, we found that weed control with WeedZap improved in shaded conditions. The greenhouse temperature was around 80 °F, so it may be that sunlight is less of a factor under warm temperatures.

Table: Weed control with weedzep (10% v/v) in relation to adjuvant, spray volume and light levels. Plants grown in the greenhouse in either open conditions or under shade cloth, which reduced light by 70%.

	Pigweed control (%)		Mustard control (%)	
	Sun	Shade	Sun	Shade
Weedzap+0.1% v/v Eco silwet (10 gpa)	31.7	93.3	26.7	35.0
Weedzap+0.5% v/v Eco silwet (10 gpa)	31.7	48.3	43.3	71.7
Weedzap+0.5% v/v Natural wet (70 gpa)	26.7	94.7	26.7	30.0
Untreated	0.0	0.0	00	0.0
LSD.05*	5.7		11.5	
*Values for comparing any two means. Pigweed and mustard were each analysed separately.				

Economic Considerations

Organic herbicides all work if you have enough volume and concentration to directly contact the weeds. However, these herbicides are expensive and may not be affordable for commercial crop production at this time. Cost in 2010 was about $400 to $600 an acre for broadcast application, which may be considerably more expensive than hand weeding. Moreover, because these materials lack residual activity, repeat applications will be needed to control perennial weeds or new flushes of weed seedlings. We see these herbicides eventually being used commercially with camera-based precision applicators that "see" weeds and deliver herbicides only to the weeds, not to the crop or bare ground.

Cover Crops for Suppression of Weeds

A growing cover crop can suppress weeds in several ways:

- Direct competition.

- Allelopathy—the release of plant growth–inhibiting substances.

- Blocking stimuli for weed seed germination.

- Altering soil microbial communities to put certain weeds at a disadvantage.

After a cover crop is tilled in, mowed, rolled, or otherwise terminated, its residues can prolong weed suppression by:

- Physically hindering seedling emergence (if residues are left on the surface as mulch).

- Releasing allelopathic substances during decomposition.

- Promoting fungi that are pathogenic to weed seedlings.

- Tying up nitrogen (N) (when low-N residues are incorporated into soil).

Competition

A vigorous, fast-growing cover crop competes strongly with weeds for space, light, nutrients, and moisture, and can thereby reduce weed growth by 80–100% for the duration of the cover crop's life cycle. Timely cover crop plantings occupy the empty niches that occur in vegetable production systems:

- After vegetable harvest.

- Over winter.

- Before planting a late-spring or summer vegetable.

- Between wide-spaced rows of an established crop.

Buckwheat, soybean, and cowpea planted in warm soil can cover the ground within two or three weeks. This "canopy closure" puts tiny, emerging weeds in the shade and hinders their growth. Summer or winter annual grasses like sorghum–sudangrass, various millets, oats, rye, and wheat form dense, fibrous root systems that appropriate soil moisture and nutrients, leaving less for the weeds. Combining a grass with a legume or other broadleaf crop is often more effective than growing either alone.

In this mature winter cover crop, the cereal rye has permeated the topsoil with a dense fibrous root system and provided support for the hairy vetch, allowing the latter to grow more vigorously and cast dense shade on the soil surface. Very little weed biomass was found in this cover crop, photographed here in late May on Cape Cod, MA.

Figure: This buckwheat (left), planted immediately after a vegetable harvest, has nearly covered the ground within 15 days after planting (DAP). Pearl millet (right) has formed substantial biomass by 42 DAP and effectively crowded out most weeds

Figure: A cover crop biculture of grass–legume can compete more effectively against weeds than either component alone

Fast-growing millets, forage soybeans, and sorghum–sudangrass can attain heights of four to seven feet, and aboveground dry biomass of four tons per acre within 65–70 days after planting (DAP). The grasses can mop up 100–150 lb N per acre in that time, and soybeans can fix up to 200 lb N per acre. Winter cereal grains, especially rye, can grow at temperatures just a few degrees above freezing, and thereby get a jump on early spring weeds. Oats and field peas planted in early spring can reach three to four feet and three tons per acre by the summer solstice.

Clovers get off to a slow start and are not initially good competitors. However, clover seedlings, especially red clover, are quite shade-tolerant; thus clovers can be interplanted or overseeded into standing vegetable crops. When the vegetable is harvested and cleared off, the established clover seedlings grow rapidly, and taller varieties—such as mammoth red, crimson, berseem, and ladino clovers—can compete well against postharvest weeds.

Competition from a strong cover crop can virtually shut down the growth of many annual weeds emerging from seed. Perennial weeds that emerge or regenerate from roots, rhizomes, or tubers are more difficult to suppress, but even their growth and reproduction can be substantially reduced by the most aggressive cover crops.

As long as the cover crop is actively growing, intercepting light, and utilizing soil moisture and nutrients, later-emerging weeds have little opportunity to grow. Tilling the cover crop into the soil as a green manure terminates the competitive effect, leaving an open niche which should be occupied by planting a subsequent crop as soon as practical.

Allelopathy

All plants give off various substances that can affect the growth of other plants. Active compounds may be exuded by living plant roots, washed off the leaves and shoots into the soil by rainfall, or released from decaying residues. These allelochemicals, some of which are potent enough to be considered nature's herbicides, have the greatest impact on germinating seeds, seedlings, and young plants, retarding their growth, causing visible damage to roots or shoots, or even killing them outright. Allelopathic effects strong enough to contribute significantly to weed control in field conditions have been documented for rye and other winter cereal grains, sorghum and sorghum–sudangrass hybrids, lablab bean, rapeseed, buckwheat, and subterranean clover.

Cover crops in the brassica family, including rapeseed, mustards, and radishes, contain a number of compounds called glucosinolates, which break down into powerful volatile allelochemicals called isothiocyanates during residue decomposition, which can affect plant growth as well as microbial activity. In field trials, some brassica cover crops have suppressed weed growth for several weeks or months after the cover crop was tilled in. However, it has been shown that weed suppression by radish cover crops is primarily a result of light exclusion that inhibits weed germination, not allelopathy.

Because each plant species gives off a unique combination of potentially allelopathic substances, and is itself sensitive to some allelochemicals and tolerant to others, allelopathic interactions are often species specific. For example, winter rye and its residues are quite active against pigweeds, lambsquarters, purslane, and crabgrass, and far less so against ragweeds, sicklepod, and morning glories. Sunflower and subclover suppress morning glories, ande sorghum can inhibit purple nutsedge and Bermuda grass as well as many small-seeded annuals.

Cover crop allelopathy can hurt some vegetables as well, particularly small seeded crops that are direct sown too soon after the cover crop. Lettuce seedlings are especially sensitive to allelochemicals, while large-seeded and transplanted vegetables are generally more tolerant. Tomatoes and other solanaceous vegetables thrive when transplanted through recently-killed residues of rye and hairy vetch. Winter grain cover crop residues have been reported to reduce growth of cabbage, but to stimulate peas, beans, and cucumbers.

Unlike direct competition, allelopathic weed suppression can persist for a few weeks after a cover crop is terminated. Tilling the top growth in as a green manure causes an intense but relatively brief burst of allelopathic activity throughout the till depth. Leaving the residues on the surface as an in situ mulch creates a shallow (less than one inch) but more persistent allelopathic zone that can last for three to ten weeks depending on weather conditions. Thus no-till cover crop management offers a potential for selective suppression of small-seeded annual weeds in transplanted and large-seeded vegetables, whose roots grow mostly below the allelopathic zone.

In addition to this "selectivity by position," some allelochemicals may be inherently selective toward larger seeds. In petri dish germination tests, green pea seeds (large) were far more tolerant to low (1–5 ppm) concentrations of various isothiocyanates than redroot pigweed seeds (small), with barnyard grass seeds (medium) showing intermediate sensitivity. Similar selectivity has been observed in field studies, on vegetables grown after brassica cover crops. Whereas the weed suppressive effects of the cover crops persisted for at least part of the vegetable growing season, yields were either unaffected or improved in potatoes, peas, spinach (direct-sown), onions (from sets), and transplanted lettuce.

Weed Seed Germination

While a brief flash of unfiltered daylight, or even a few minutes of full moonlight, can trigger germination of many small-seeded weeds, the green light that reaches the soil beneath a closed canopy of plant foliage tends to inhibit germination. This is because many seeds sense the quality of light by means of a special compound called phytochrome that works as a molecular switch. Red light (abundant in daylight) flips the switch to "germinate now" whereas light that is poor in red and rich in far-red (a wavelength between red and infrared, barely visible to the human eye) flips the switch to "go dormant". The chlorophyll in green leaves absorbs most of the red light and transmits the far-red, and the phytochrome in weed seeds senses the filtered light as a signal that a shading canopy is present, rendering conditions unfavorable to weed growth. Many early spring annual weeds initiate germination in fall, and the startling spring weed suppression after a daikon radish crop is primarily a result of changes to the light quality created by a completely closed radish canopy. This makes early cover crop seeding and complete canopy closure essential for optimizing weed suppression by forage radish. Part of the weed-suppressive effects of hairy vetch cover crops have also been attributed to this light quality effect, and this phenomenon may also contribute to the weed suppression sometimes observed after other dense-canopy cover crops like buckwheat.

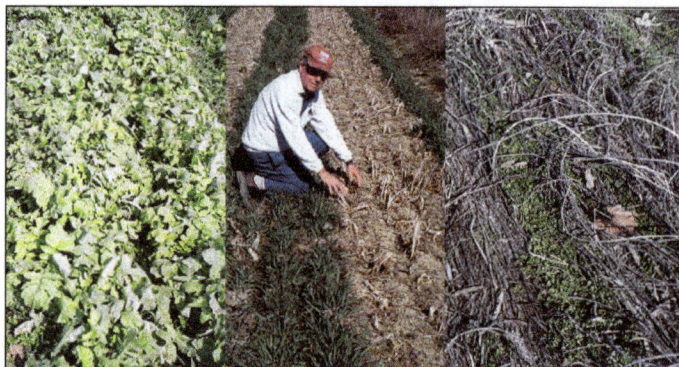

Figure: A daikon radish cover crop, sown in August, covered the ground with a heavy canopy by midautumn (left)

The crop winter-killed and its residues mostly disappeared by March, yet Professor Ron Morse of Virginia Tech could find almost no winter weeds in the radish plots (center), whereas common chickweed grew vigorously through the more persistent residues of other winter-killed cover crops (right). Early spring chickweed and many other annual weeds initiate germination in fall. By intercepting light, the fall radish canopy prevents weed seed germination.

Several field studies have documented a decline in annual weed populations in cultivated fields that are rotated to red clover for one or more years. With few or no annual weeds growing and replenishing the weed seed bank, weed seed numbers decline through seed predation, physiological aging, and decay.

Figure: This year-old stand of red clover casts dense shade and alters the quality of light reaching the ground so that seeds of most annual weeds are no longer stimulated to germinate.

Effects on Soil Microbial Communities

Each plant species exudes through its roots a characteristic mix of substances, including carbohydrates, amino acids, organic acids, and other "microbial food", as well as its particular set of allelochemicals. This biochemical mix elicits and supports a specific microflora (community of fungi, bacteria, protozoa, and other microorganisms) in the plant's rhizosphere (the soil immediately adjacent to the plant roots); to a lesser degree, it also influences the microflora of the bulk soil. The microbes fostered by one plant species can help, hinder, or even sicken another plant species.

A vigorous cover crop with an extensive root system that harbors microorganisms harmful to certain weeds can thereby provide an added measure of control of those weeds. For example, most grain and legume cover crops are strong hosts for mycorrhizal fungi which live as root symbionts and enhance crop growth. Several major weeds, including pigweeds, lambsquarters, nutsedges, purslane, and weeds in the buckwheat family, are nonhosts that do not benefit from mycorrhizae, and may exhibit reduced vigor if their roots are invaded by mycorrhizal fungi. Several researchers have begun to explore the potential of mycorrhizal fungi as a weed management tool.

Plant root exudates and plant-microbe interactions can also influence certain species or classes of microorganisms in the soil as a whole, with subsequent effects on other plants. For example, the glucosinolates and isothiocyanates released by crops and weeds in the crucifer family (such as brassica crops, wild mustards, and yellow rocket) can inhibit soil fungi, including some pathogens. Crucifers and other nonmycorrhizal host plants, while not directly toxic to mycorrhizae, do not support the high populations of active mycorrhizal fungi often found in the soil after strong-host species such as most legumes.

Crop–weed–soil–microbe interactions are one of the cutting edges in organic weed management research. Scientists are searching for specific microbial species or floras that thrive in the root zone of widely-used cover crops, and that attack or suppress major weed species without posing a serious threat to the desired vegetable crops. These relationships are complex, and practical applications are some years or decades away.

Mulch Effect

When a cover crop is killed by temperature extremes, mowing, or rolling, residues left on the soil surface as a mulch can continue to hinder weed growth for some time. By keeping the soil surface shaded and cool, and by reducing daily fluctuations in soil temperature, the organic mulch reduces the number of weed seeds that are triggered to germinate. Small-seeded broadleaf weeds that do sprout are often effectively blocked by a 2–3 inch thick layer of cover crop residues. Larger-seeded broadleaf seedlings, grass seedlings, and perennial weed shoots from buried rhizomes and tubers will eventually get through, though even their growth may be delayed by residues of a high biomass cover crop.

The mulch effect can be enhanced by the release of allelopathic substances from the decaying residues, as noted earlier. In addition, organic mulch provides habitat for ground beetles and other predators of weed seeds, as well as microorganisms that can attack and kill weed seedlings.

Weed suppression by cover crop residue can vary from negligible to highly effective for anywhere from two weeks to several months, depending on cover crop biomass and nitrogen (N) content, season, weather, and soil conditions. Warm, moist weather combined with high soil biological activity accelerates decomposition of cover crop residues and their allelochemicals, thus shortening the weed control period. Strawy, low-N residues last longer than succulent, high-N residues. In dry climates, the weed suppressive effect of even a legume cover crop mulch can be substantial.

Figure: This rye–vetch cover crop mulch delayed weed growth sufficiently to prevent significant weed competition against the broccoli

The mulch effect effectively blocked most annual weeds, while a few perennial quack grass are beginning to break through. The cover crop was mowed and the broccoli transplanted about seven weeks before this picture was taken on Cape Cod, MA.

Green Manure Effects

Tilling a cover crop into the soil as a green manure stimulates a flush of microbial activity that can make the soil temporarily inhospitable to most weeds and crops. The tillage itself stimulates weed seed germination, but the incorporated residues may promote damping-off fungi and other pathogens that then attack the weed seedlings. If the residues are rich in carbon (C) relative to N (C:N ratios of 30 or higher), soil microbes will immobilize (tie up) plant-available soil N while consuming the C-rich organic matter, and thereby slow the growth of weed seedlings. These effects—combined with the brief intense flush of allelochemicals from certain cover crops, especially radish and other brassicas—can help clean up a weedy field.

On the other hand, leguminous or young, succulent green manures provide plenty of N and other nutrients that can stimulate a burst of weed emergence and growth, thereby negating earlier weed-suppressive effects of the cover crop.

Figure: A farmer on Cape Cod, MA plows down a winter cover crop of hairy vetch in late spring

The succulent, high-nitrogen legume cover crop will decompose rapidly and require only a short (one to two week) waiting period before vegetables can be planted. The disadvantage to this practice is that it may also open a highly fertile niche for weed growth.

Note that cash crops are also subject to green manure effects. Vegetables should not be planted during the microbial flush after soil incorporation of a green manure. Careful timing is essential to avoid adverse effects of green manure on vegetables, yet take advantage of temporary weed-suppressive effects that can give the vegetable a head start on the weeds.

Ecological Weed Management in Organic Vegetables

Ecological weed management begins with careful planning of the cropping system to minimize weed problems, and seeks to utilize biological and ecological processes in the field and throughout the farm ecosystem to give crops the advantage over weeds. In addition, mechanical and other control measures are usually needed to protect organic crops from the adverse effects of weeds. This is particularly true in vegetables and other annual crops, for which production practices keep natural plant succession at its earliest stages, thereby eliciting the emergence of pioneer plants that can become agricultural weeds.

While tillage and cultivation can degrade soil quality and increase the risk of erosion losses, many other organic weed management tools are more soil-friendly. For example, a diversified rotation of vigorous cash crops and cover crops can enhance soil organic matter, tilth, and fertility, provided that a sufficient quantity and diversity of residues are returned to the soil to feed the soil life. Grazing livestock after a production crop to remove weeds or interdict weed seed set can add fertility in the form of manure, though intensive grazing can also compact the soil. In the interest of food safety, care must be taken to avoid direct contact of fresh manure with vegetables and other food crop. Mowing and flame weeding (if properly done to avoid excessive heating of the soil itself) are much easier on soil structure than cultivation, and can be just as effective in certain stages of weed and crop development. Mowing or rolling a cover crop to form an in situ mulch can enhance the soil benefits of the cover crop, compared to tilling it in, and can effectively suppress many annual weeds. Other organic mulches, such as straw and chipped brush, add organic matter, whereas synthetic clear or colored plastic films and weed barrier fabrics do not. All mulches are very effective in preventing soil erosion.

Table: A summary of organic weed management tools.

	Preventive	Control
Major tools:		
The Grower's Mind (planning, observation, and ingenuity)	X	X
Vigorous Cash Crops	X	
Crop Rotation	X	
Cover Crops	X	
Organic Mulches	X	X
Opaque Synthetic Mulches (black plastic, etc.)	X	X
Conservation Biological Control (conserve weed consumers present on farm)	X	X
Livestock	X	X
Tillage and Cultivation Tools and Implements		X
Mowers and other Cutting Tools		X
Rollers and Roll-crimpers (for converting mature cover crops into in-situ mulch)	X	X
Flame Weeders		X
Minor and experimental tools:		
OMRI certified organic herbicides		X
Bioherbicides (specific pathogens of weeds)		X
Management of soil microflora	X	X
Specific crop–weed allelopathic interactions	X	X
Classical biological controls for specific weeds (usually against invasive exotic weeds in rangeland and natural ecosystems)		X
Clear plastic mulch (soil solarization)	X	X

Ecological weed management consists of many-component strategies tailored to each region, cropping system, and farm. Matt Liebman and Eric Gallandt describe the process as using "many little hammers", including "indirect controls", such as crop variety, planting date, and nutrient management, rather than relying only on the "direct controls" or "large hammers" of cultivation and herbicides. In their words, "the use of a combination of methods can lead to (i) acceptable control through the additive, synergistic, or cumulative action of tactics that may not be effective when used alone, (ii) reduced risk of crop failure or serious loss by spreading the burden of protection across several methods, and (iii) minimal exposure to any one tactic, and consequently reduced rates at which pests adapt and become resistant."

Steps do not comprise a precise linear sequence of instructions; rather they offer a conceptual framework within which each farmer can develop a site-specific strategy. This process requires systems thinking, which views the field as a complex system of interacting components—such as crops, weeds, soil, insects, and microorganisms—that form a web of relationships, not a linear sequence of cause-and-effect. Similarly, the following steps are employed together in a synergistic manner, and thus differ from the sequence of instructions for assembling a car or a farm implement. For example, Step 6 (cover crops) can be seen as a part of Step 2 (minimize niches for weeds), and Step 1 (know the weeds) provides vital information for other steps, particularly steps 3 (keep the weeds guessing), 4 (design for effective weed control), and 7 (manage the weed seed bank). Biological processes (Step 9) include indigenous biocontrols that help reduce the weed seed bank (Step 7) as well as the competitive and allelopathic effects of cover crops (Step 6). Step 11 (observe weeds and adapt practices) is an ongoing feedback loop that informs and fine-tunes all the other steps. Utilizing this or another suitable framework, the organic grower selects and assembles a set of "many little hammers" that, working together, keep the farm's weeds from becoming major weed problems.

Planning Steps

1. Know the Weeds Obtain correct identification of the major weeds present on the farm. Monitor fields regularly throughout the season. Keep records on what weeds emerge at different seasons, and on efficacy of any preventive and control measures taken. Learn each weed's life cycle, growth habit, seasonal pattern of development and flowering, modes of reproduction and dispersal, seed dormancy and germination, and how the weed affects crop production. Find the weed's weak points—possibly the stages in its life cycle that are most vulnerable to control tactics—and stresses to which the weed is sensitive; these can be exploited in designing a management strategy.

"Know the weeds" is listed first because it informs most of the succeeding steps. However, gaining a thorough knowledge of the farm's weed flora is an ongoing process over many seasons (perhaps the lifetime of the farmer!) that drives the year-to-year refinement of the farm's weed management system.

2. Design the Cropping System to Minimize Niches for Weed Growth In planning the crop rotation, avoid creating open niches in time or space. Plan tight rotations that follow one crop harvest promptly with the next planting. Open niches in space between crop rows can be reduced by using a narrower row spacing, intercropping, relay cropping, overseeding cover crops into established vegetables, or no-till management of cover crops prior to transplanting vegetables.

3. Keep the Weeds Guessing with Crop Rotations Plan and implement diversified crop rotations that vary timing, depth, frequency, and methods of tillage; timing and methods of planting, cultivation, and harvest; as well as crop plant family. Alternate warm- and cool-season vegetables. Rotate vegetable fields into perennial cover for two or three years to interrupt life cycles of annual weeds adapted to frequent tillage. Schedule tillage and cultivation operations when they will do the most damage to the major weed species.

4. Design the Cropping System and Select Tools for Effective Weed Control Develop control strategies to address anticipated weed pressures in each of the farm's major crops. Choose the best cultivation implements and other tools for cost-effective preplant, between-row, and within-row weed removal. Plan bed layout, as well as row- and plant spacing, to facilitate precision cultivation. Choose irrigation methods and other cultural practices that are compatible with planned weed control operations.

Preventive Steps During the Season

1. Grow Vigorous, Weed-competitive Crops A healthy, fast-growing crop that can outcompete weeds is the best way to prevent weed problems. Choose locally-adapted crop varieties that grow tall or form lots of foliage that can shade out weeds. Maintain healthy, living soil. Provide optimum growing conditions—planting date and spacing, moisture, soil tilth and aeration, fertility, and pest

and disease management. Deliver water and fertilizer within-row to feed the crop and not the weeds. Note that either insufficient or excessive levels of major nutrients (nitrogen, phosphorus, and potassium) can give certain weeds a competitive advantage over the crops.

2. Put the Weeds Out of Work—Grow Cover Crops! Cover crops do the same job as weeds, only better. They rapidly occupy open niches, protect and restore the soil, provide beneficial habitat, add organic matter, and hold and recycle soil nutrients. They suppress weeds through direct competition and sometimes through allelopathy—the release of plant-growth-inhibiting substances into the soil. Whenever a bed or field becomes vacant, plant a cover crop immediately so that it can begin the vital restorative work that nature accomplishes with pioneer plants or weeds. Good cover cropping plays a major role in Step 2 (minimzing open niches), and can put the weeds out of a job.

3. Manage the Weed Seed Bank—Minimize "Deposits" and Maximize "Withdrawals" Prevent formation and release of viable weed seeds, and proliferation of rhizomes and other propagules of perennial weeds. Avoid importing new weeds with manure, mulch hay, and other materials from off-farm sources. Utilize stale seedbed, cultivated fallow, or targeted tillage practices to draw down seed banks of the major weeds present. Encourage weed seed mortality and weed seed consumption by ground beetles and other organisms.

Control Steps During the Season

1. Knock Out Weeds at Critical Times Plant vegetables into a clean seedbed, hit early-season weeds while the are small, and keep crops clean through their critical weed free period (through the first third or half of the life cycle of most vegetables). Prevent seed set by "escapes" and late season weeds. When practical, interrupt vegetative propagation by invasive perennial weeds through timely removal of top growth.

2. Utilize Biological Control Processes to Further Reduce Weed Pressure Rotate livestock, poultry, or weeder geese through fields to graze weeds and interrupt seed set. To ensure food safety and comply with USDA Organic Standards, time such grazing so that fresh droppings are not deposited any less than 120 days prior to harvest of the next crop. Encourage weed seed predation and decay by maintaining high soil biological activity and providing habitat (mulch, cover crops, hedgerows) for belowground and aboveground weed seed consumers (conservation biological control). Enhance overall soil biological activity to tip the competitive balance in favor of crops, and possibly to shorten the "half life" of the weed seed bank.

Classical biological controls (introduced natural enemies) are commercially available for a few invasive exotic weeds.

3. Bring Existing Weeds Under Control Before Planting Weed-sensitive Crops Weed control in perennial horticultural crops like asparagus, small fruit, and some cut flowers can be quite difficult, especially when perennial weeds dominate the weed flora. Bring existing weed pressures under good control through repeated tillage and intensive cover cropping before planting any perennial vegetable, fruit, or ornamental crops. Choose fields with the best weed control or lowest weed pressure for weed-sensitive annual vegetables with a long critical weed free period, such as carrot, onion,and parsnip. Be sure weeds, especially perennial weeds, are under good control before attempting no-till management of cover crops prior to cash crop planting.

Enhancing the Organic Weed Management System – Observe, Adapt and Experiment

Keep Observing the Weeds and Adapt Practices Accordingly Note and record any changes in weed species composition, emergence and growth pattern, or weed pressure, and modify practices as needed. For example, an increase in certain annual "weeds of cultivation" may indicate a need to reduce tillage or diversify the crop rotation. An increase in invasive perennials may require tilling deeper or more aggressively for a time. Watch out for the arrival of new weed species that could pose problems.

Expect weed populations and flora to shift over time. Every farm decision and field operation can elicit changes in the weed community, as can weather variations, to say nothing of long term climate changes. "Reading" the weeds each year becomes an information feedback loop, guiding weed management practices for the following season.

An effective organic weed management system cannot be spelled out precisely because ecological weed management is inherently site specific and responsive to changes in the farm ecosystem. There is effectively no "organic weed control cookbook" to replace the precise herbicide protocols that have been developed for conventional production of row crops and some vegetables. No scientist can come up with a better weed management strategy for a particular farm than the strategy a skill full organic farmer can develop by applying ecological weed management principles to the particular suite of crops, weeds, soil conditions, and available resources on her or his farm.

References

- Insect-pest-management-in-organic-farming-system: intechopen.com, Retrieved 23 June, 2019
- Colucci-Organic-Disease-Management.: carolinafarmstewards.org, Retrieved 15 May, 2019
- Biological-pest-control, entry: newworldencyclopedia.org, Retrieved 12 April, 2019
- Biological-control-of-insect-pests: extension.org, Retrieved 21 March, 2019
- Weed-management: gardenorganic.org.uk, Retrieved 13 July, 2019
- Direct-weed-controls: gardenorganic.org.uk, Retrieved 16 June, 2019
- Mulching-for-weed-management-in-organic-vegetable-production: extension.org, Retrieved 17 February, 2019
- Organic-mulching-materials-for-weed-management: extension.org, Retrieved 21 April, 2019
- Cultural-weed-controls: gardenorganic.org.uk, Retrieved 13 July, 2019
- What-is-organic-no-till-and-is-it-practical: extension.org, Retrieved 1 August, 2019
- How-cover-crops-suppress-weeds: extension.org, Retrieved 28 April, 2019
- Twelve-steps-toward-ecological-weed-management-in-organic-vegetables: extension.org, Retrieved 19 May, 2019

Chapter 4

Organic Fertilizers

The fertilizers that are derived from organic sources such as human excreta, animal matter, animal excreta and vegetable matter are known as organic fertilizers. Peat, manure and animal waste are some of the naturally occurring organic fertilizers. This chapter has been carefully written to provide an easy understanding of the varied facets of organic fertilizers as well as the processes used in their production, such as vermicomposting.

Organic fertilizers are made of all the natural products like animal matter, animal manure, vegetable matter like compost and crop residues. And organic fertilizers not only add required nutrients for the plants, but also promote fertility of the soil by increasing the solid structure and draining system of the soil.

Advantages and Disadvantages of using Organic Fertilizer

Benefits of using Organic Options

Until early in the twentieth century, composted manure, kitchen scraps, and other organic wastes represented the only means of improving soil fertility. Organic farming and gardening were not moral or environmental choices, but simply the way of life. Organic fertilizers are an all-encompassing method that build soil structure and composition as they slowly release nutrition to the plants.

Organic fertilizers have all the natural ingredients, organic fertilizers release nutrients slowly, provides a steady flow of plant nutrients. Organic nutrients are non-toxic, they don't harm roots. Organic fertilizer improves soil structures, ground water, fertility, and promotes earthworms and microorganism in the soil. Improves soil fertility time to time and balances chemical imbalances permanently for a long time.

Work of Chemical Fertilizers

Chemical fertilizers work on the soil and plants instantly. Chemical fertilizer is toxic, they harm the roots. Chemical fertilizers destroy the organic matters in the soil. Continuous application of fertilizers will kill small worms and microorganism in the soil that promotes the draining system of the soil. Salts in the chemical fertilizers will make the soil more acidic and kills earthworms.

Organic Fertilizers

Organic fertilizers improve garden soil time to time. The make the soil loose and promotes air movement to the roots the plants. There are many types of organic fertilizers that can fast and slow impact on the growth of the plants. Organic fertilizers are extracted from animal materials, plants and natural mineral rocks.

Types of Organic Fertilizers

Dry Organic Fertilizers

Dry organic fertilizers are slow release fertilizers, they are mixed into the soil directly. you can use the dry fertilizers in ground garden soils and potting mixes. This type of organic fertilizers is used for long term seedlings, transplants and crops. Dry organic fertilizers are usually added during the plantation, they are mixed into the soil.

Liquid Organic Fertilizers

These fertilizers are in the form of liquids, the provide light and instant energy to the plants during the growing season. These types of liquid fertilizer are easily absorbed into the soil. These fertilizers are poured around the soil of the plant, that can be easily absorbed by the roots. Liquid fertilizers may be sprayed on the foliage also, these are mainly used for vegetables and fruit plants during the growing season. Feeding the leaf's can supply the plant with the necessary nutrients that are not available in the soil or if the roots are stressed. Liquid fertilizers are can be used forever two weeks or every month during growing season and fruiting period. Liquid fertilizers are applied in the early morning or evening, the fertilizers are easily and quickly absorbed by the plant without harming the foliage. Liquid fertilizers are sprayed till the liquid starts dripping from the leaves.

Growth Enhancers

Growth enhancers are not fertilizers, they just boost up the power of plants to absorb the nutrients from the soil. Kelp is the one of common growth enhancer, used by the farmers. Growth enhancers come in both the liquid form and dry form. Growth enhancers has enzymes and growth promoting hormones. Growth enhancers have capacity to stimulate the bacteria in the soil that increases the soil fertility. Make soil loose and well drained.

Use of Organic Fertilizers

Organic fertilizers are used same as the chemical fertilizers, just buying pre-made supplies and apply as per the instructions. Or if you want to prepare yourself as per our garden need, you can do following an exact procedure. Don't overdo it to avoid root burn or it can kill sensitive plants.

Ways to Apply Dry Organic Fertilizers

Follow the labeled instructions before using fertilizer. Dry organic fertilizers are applied before planting, just add a 1-inch layer of dry organic fertilizer to the soil. Now rake the soil, so that fertilizer reaches up to 4 to 6 inches in the soil. And add some fertilizers to planting holes or rows. Side dressing plants with dry organic fertilizer's during growing season also provide necessary nutrients to the plant.

Ways to Apply Liquid Organic Fertilizers

Follow the labelled instructions. Liquid fertilizer is applied only during morning and evening times. Liquid fertilizers may be sprayed on the foliage or can be watered around the plant. Sprayer used should have a fine mist.

Ways to Apply Growth Nutrients

Follow the complete label instructions when you are applying growth enhancers as foliar sprays to the plant leaves. Adding growth boosters to the soil during planting, will increase the soil fertility makes soil well drained. Growth enhancers can be given once in a month for five months of growing seasons depending on the crop growing period.

List of Organic Fertilizers and Types of Organic Fertilizers

Types of Organic Fertilizers – Plant Based Organic Fertilizers: These Organic fertilizers are made from Plants.

Alfalfa Meal

Alfalfa meal is a plant based organic fertilizer made from fermented alfalfa plants. This fertilizer has low amounts of nitrogen, phosphorous and potassium. It comes in a powdered form and dissolves in the soil very easily. This organic fertilizer is used to rebuild soil structure and organic matter by providing nutrients to the roots. It is a moderate release fertilizer acts as conditioner to the soil for early spring crops. N: P: K ration in Alfalfa meal is 3:2:2.

Cottonseed Meal

Cottonseed meal is a plant based organic fertilizers, it is a rich source of nitrogen and low amounts of phosphorous and potassium. It is a slow release fertilizer mainly used of conditioning the garden soils before you cover the crops or before the mulch is applied. Cottonseed meal is a popular fertilizer that promotes the healthy growth and beauty of grasses, ornamental plants and vegetables. N: P: K ratio in cottonseed meal is 6:2:1.

Corn Gluten Meal

The Corn Gluten meal is a slow release plant based organic fertilizer works as a good stabilizer for the soil. It is a dry organic fertilizer, and it takes time to break down into the soil and it can break down over winters. It is applied early spring and again in autumn. Corn gluten meal is a bi-product of corn (maize) and it can control weeds formation. N: P: K ratio in Corn gluten meal is 0.5:0.5 :1. Corn gluten meal can be applied 5 to 6 weeks before sowing or two weeks after sowing.

Composts

Composts are plant based organic fertilizers. Composts are decomposed form of organic matter, and the process of decomposing is also called composting. Composts are rich in nutrients and used for gardens, landscaping, organic agriculture, container gardens, backyard gardening, etc. Composts act as a good soil conditioner and boosts up the organic matters and the fertility of the soil. Compost provides a rich growing medium for the plants that can hold moisture for a long time and provide necessary nutrients for the plants. Compost can be used in many ways, it can be mixed in the soil or can be used as a mulch. Compost tea is used as a foliar spray. Compost side dressing around plants will give plants a constant supply of nutrients. Compost should have added before or after planting, it acts as a good soil conditioner between growing seasons. N: P: K levels in compost are 2:1.5:1.5. Types of Organic Fertilizers – Soybean Meal: Soybean Meal is plant based organic fertilizer, it is a rich source of nitrogen and phosphate and neutral pH levels. Soybean meal is dry type organic fertilizer, works at a moderate speed. It is used as a long-term soil conditioner. Soybean meal is added once or twice during the growing season of the plants, it improves the leaf growth. Application of soybean meal to plants should be done as per the labelled instructions. N: P: K ratio in soybean meal is 7:2:0.

Kelp Fertilizers

Kelp is a plant based organic fertilizers, made from a sea plant. Kelp is a rich source of trace elements and a good source of hormones that helps plants to grow up to full potential. It is a liquid organic fertilizer used by mixing with water and used as a foliar spray. N: P: K ratio of kelp fertilizer is 1:0.2:2.

Seaweed Fertilizers

Seaweed fertilizer is a plant-based fertilizer, it is instant release fertilizer. It contains all the major nutrients in minimum quantities and container zinc, copper and iron in large quantities. Seaweed is the best fertilizer for grain-based crops that need high levels of potassium. The N: P: K levels in seaweed is 1.5:0.75:5.

Glass clippings

It is a plant based organic fertilizers, it is mainly used to prevent weeds and control the moisture levels in the soil. It used as mulch, adding a 2-inch layer of grass clippings is needed for a growing season. N: P: K ratio in grass clippings is 1:0:1.2.

Mineral based Organic Fertilizers

Organic fertilizers derived from natural occurred minerals are called mineral based organic fertilizers.

Rock Phosphate

This organic fertilizer is extracted from the mineral rocks and clay. It is a good source of phosphate and other essential nutrients. Rock phosphate is very slow release fertilizers, it dissolves in water and stands around the soil till the plants absorbs it. It is used to increase the acidic nature of the soil. Rock phosphate promotes growth of the transplants and seedling. Rock phosphate is highly acidic, before using it check pH levels of the soil. The N: P: K ratio in rock phosphate is 0:5:0.

Green Sand

Green sand, organic fertilizer is a rich source of iron, potassium and magnesium. Green sand promotes the better growth of the plant, it loosens the soil, improves moisture levels in the soil and softens the ground water and improves the root establishments. N: P: K ratio in green sand is 1:1:5, nutrient levels in the green sand depends upon the on their source.

Animal based Organic Fertilizers

Organic fertilizers made from the animal's blood, excreta and animal matter.

Cow Manure or Cow Dung

Cow manure is highly used animal based organic fertilizer for its good pack of nutrients. Cow manure can control weeds, and increases the moisture holding capacity of the soil and increases air penetration in the soil. T Beneficial bacteria in cow dung support the growth of young roots by releasing nutrients slowly in easily accessible forms. Cow manure should be mixed with lighter material like straw or hay or vegetable matter or garden debris. It is used as a top dressing for the soil in gardens, containers, crops and it acts in a moderate speed. N: P: K ratio in cow dung is 2.5:1:2.5. too much use of cow manure can harm and burn the plants.

Livestock Manure

Poultry manure like poultry crap is a nutrient rich animal based organic fertilizer that works very fast. It is mainly used for soil low in nitrogen levels. The livestock manure can be used after harvesting crops or before the beginning of gardening cycles. Early spring and early fall season are the best season to apply poultry manure. N: P: K ratio in poultry manure is 3.5:1.5:1:5.

Earthworm Casting

It is an animal based organic fertilizers, it is a bi-product produced from a composting process using earthworms. This is also called vermicomposting, contains water soluble nutrients and a nutrition rich organic fertilizer and soil conditioner, it is can used for flower and vegetable gardens. N: P: K ratio in earthworm castings is 2:1:1.

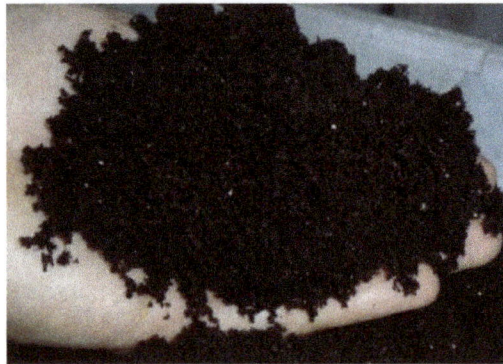

Vermicompost Biohumus Fertilizer

Blood Meal

Blood meal is an animal based organic fertilizer, it is a dried form of animal blood. Blood meal raises the nitrogen levels in the soil and makes plants grow bushy and greenish. Blood meal releases nitrogen rapidly, promotes flowering and fruiting's and used as a natural pest repellent. It can be added dry or water dissolves to the roots of the plants as per the labelled instructions and best to add blood meal to the soil before planting. The highly acidic nature of blood meal can harm young plants if used in large quantities. N: P: K levels in blood meal is 12:1.5:0.5.

Bone Meal

Bone meal is an animal based dry organic fertilizer, is a form of grounded form of animal bones. Bone meal is a slow release fertilizer used as source of phosphorous, calcium and other proteins. Bone meal promotes growth of the transplants and seedling. Phosphorous in bone meal will help flowers and fruits to grow bigger and increases the yields. N: P: K ratio in bone meal is 4:20:0.

Feather meal

Feather Meal is animal based organic fertilizers, is a grounded form of poultry feathers. Feathers meal is a rich source of nitrogen and it doesn't contain any calcium or phosphorous. It added prior to planting to boost up the nutrients in the soil or side dressed around the plant for a constant supply of nitrogen. N: P: K ratio in feather meal is 12:0:0.

Seabird Guano

It is animal based organic fertilizers. It is a suitable fertilizer for plants and lawns and make plants greenish and healthy. Seabird guano acts as a natural fungicide and control nematodes in the soils. It is the best organic fertilizer, with decent amounts of nitrogen, phosphorous, calcium and containers good amounts of trace elements like iron, boron, copper etc. that support growth of plants.

Bat Guano

Bat Guano is an animal-based fertilizer, and is a rich source of nutrients like nitrogen, phosphorous, calcium, and other micro-nutrients necessary for plant growth. It is a water-soluble fertilizer used as a foliar spray. It promotes root growth and supports flowering and fruiting's and make stems stronger. N: P: K ratio in bat guano is 10:10:2.

Fish Meal

Fish meal is an animal based organic fertilizers, it is a fast release fertilizer. It is a rich source of organic nitrogen, phosphorous and calcium. Fish meal improves soil health, increases fertility and adds primary nutrients to the soil, for plants to grow healthy. N: P: K ratio in fish meal is 5:2:2.

Fish Emulsion

It is an animal based organic fertilizer, it is a liquid fertilizer. It is a source all micronutrients and acts as conditioner to the soil. Fish emulsion can be used as root drench or as a foliar spray. N: P: K levels in fish emulsion 2:4:0 to 5:1:1. Fish emulsion is highly acidic can cause burning in plants if used in large quantities. Use them as per labelled instructions.

Shell Meal

Shell meal is animal based organic fertilizer, container good amounts of calcium, phosphorous and good amount of trace elements. Shell meal is also used as natural pesticides, it protects plants from nematodes. N: P: K ratio in shell meal is 5:2:5.

Three Steps for Soil Health

Virtually every aspect of organic gardening revolves around the health of the soil. a Thriving growing substrate revolves around maintaining the delicate symbiotic relationship between soil microorganisms and the nutrients they (and the plants) require. Organic fertilizer contributes to soil health in the following ways:

1. Increased nutrition: Natural soil is rich in organic matter that releases nutrients at a steady rate. Increasing organic matter in agricultural soil improves the soil structure, creating more air space and water retention within the soil. Quality dirt is made up of a mix of particles and substances graded by permeability (how easily air, water and roots can move through it) - granular particles/stone, clay, sand, humus/organic matter. Organic fertilizer assists microorganisms to break down organic matter while allowing a metered release of that nutrition. Organic physical additives like stone dust - and peat or coconut hulls - create good soil with a balanced permeability.

2. Reduced soil erosion: A higher proportion of organic material in the soil will also prevent soil erosion, helping to avoid the dust bowl effect seen in the 1930s. Soil structure is critical to root health. Too permeable soil, like sandy substrates allow too much air and water to move through quickly. Water carries nutrition with it, and quick drainage equals a faster loss of nutrients (this run-off is what damages ecosystems and waterways near industrial agricultural fields). Likewise, clay soils, because of compaction, hold too much water and very little air and "light," dusty earth is prone to the erosion witnessed during the infamous Dust Bowl. Since organic fertilizers are inherently "found" materials - like manure, vegetable scraps, aggregates, seaweed - they build soil structure as they hold and release nutrients to the plants.

3. Healthy Ecosystem: Organic fertilizer is rooted in a complex foundation. Organic fertilizers are not used alone- they are intertwined in a gardening process that is gentler on microorganisms and earthworms living in the soil, creating a healthy ecosystem that is assisted by careful additions of physical (mulch) and nutritional (liquid seaweed) elements. Synthetic or chemical fertilizers create "burn" or bursts of ammonia, phosphorous and nitrogen that give a fast release. These fertilizers, if applied incorrectly, hinder proper fruiting and impede soil ecosystem development. Organic gardeners want to build a healthy, sustainable and environmentally beneficial farming system that becomes self-sustaining with minimal disruption.

Slow and Steady Release of Nutrients

The slow and gradual release of nutrients is listed among both a pro and con of organic fertilizer use. As an advantage, the natural release of elements means that there is a reduced risk of nutrient burn from over-fertilization. This approach also means that applications of soil amendments are required less frequently, reducing operating cost and manual labor. With organic fertilizer, nutrient availability and uptake by plants occur at roughly an equal rate, meaning nutrients are preserved in soil and plant matter rather than leaching away with rainwater. The resulting plant growth occurs at a natural, healthy pace. This tends to produce stronger, more stable plants than those grown at an artificially accelerated rate, theoretically producing improved taste and nutritional value at the same time.

Organic is Economical

Shovel Full of Compost

Organic fertilizer is potentially a cheaper option than chemical alternatives. If you have a compost or live in a rural area, the only cost is time. Many farmers will sell manure by the truckload or even give it away if you are willing to pick it up. In urban and suburban neighborhoods, a composting unit can be cheap, effective, and unobtrusive. For a nominal upfront investment, even apartment dwellers can have their own organic worm bin composting system to feed a balcony garden.

Better for the Environment

The combined influence of increased organic matter and reduced nutrient leaching means that elements such as nitrogen and phosphorus will end up in your plants' roots instead of the local waterways. Nutrient leaching from agriculture is a major culprit in the development of algae blooms on lakes and ponds. This process, known as eutrophication, disrupts ecosystems and renders water unfit for human use.

Even organic fertilizers can impact the environment if they are not stored or used correctly (think of manure run-off into streams). Since quick release conventional fertilizer products are intended for that fast uptake - the solubility means that drainage carries extra salts off of the growing area at a higher concentration rate. These leached elements enter water supplies as contaminants. Retained chemical fertilizers negatively affect the soil ecosystem creating issues with soil acidification, compaction (or soil crumb), and a systematic destruction of the delicate microflora. Over-use of chemical fertilizers will slowly "kill" off a healthy soil structure.

Always do a soil test before adding any compounds to your garden. Soil testing is easy - simply gather your sample according to the kit's directions and send the sample off. After the initial test, you can retest annually if needed, or every few years.

Drawbacks to Organic Fertilizers

Organic fertilizer holds many advantages over chemical alternatives, but it may not be best in every situation. Conventional fertilizers can be used in a beneficial way - if applied correctly in necessary growing situations. Some potential disadvantages of organics include:

- Limited nutrient availability: The slow-and-steady approach that makes organic fertilizer perfect for most applications can pose a problem in certain situations. Organic fertilizers are bound into their structures - this is what allows for the slow break-down. The release of nutrients from organic fertilizers can be dependent on both climate and the presence of microorganisms in the soil. Damaged soils may lack the necessary biological conditions for effective composting. Severely nutrient-deprived plants needing a boost might do better initially with a readily available nutrient mixture in a liquid form.

Woman working on farm

- Labor-intensive: Organic fertilizers can be bulky, messy materials. Some would argue that working with organic fertilizer is a labor of love, but turning compost piles, moving manure, and spreading solid fertilizer are not for everyone. This also means that applying fertilizer on a large scale can be more difficult, as heavy manure or blood meal granules are less suitable for mechanical spreaders.

- Potentially pathogenic: Incomplete composting can leave certain pathogens in the organic matter. These pathogens can enter the water system or the food crops, causing human health and environmental problems.

- Expensive: Commercial organic fertilizers are often more expensive per unit than comparable chemical products.

The negative aspects of manufactured fertilizer can be mitigated and reversed by careful soil testing. Add lime and compost to soils that have been treated with fertilizers. Plant health is negatively affected by the salts and acidification of too much (or too long) use of chemical products, since they can kill off the bacteria and fungal organisms that boost roost health. Mycorrhizal fungi coexist with a plant's root structure (the roots feed the fungi) and the mycorrhizae boost the plant's health by stimulating its immune system and assisting in its nutrient and water uptake.

Weighing the Choices

Depending on your situation, you will likely come up with a workable balance. Cost, convenience or the physical aspects of handling organic fertilizer, might be a determining factor for some, while the environmental and safety benefits for you and the soil will need to be factored into the process. Whatever you decide, familiarize yourself with soil health and run additional soil tests every few years. Applying the minimum amount of fertilizer (organic or chemical) necessary, and only doing so at the optimal time, will offer the best overall return for your time, money, and labor.

Organic Liquid Fertilizer

Plants absorb nutrients through both their roots and leaves, using liquid fertilizers as foliage feeding or soil drench are good ways to supply fast-growing plants like vegetables plant ready nutrients whenever needed.

There many liquid fertilizers available for the home garden, both commercially prepared and homemade. Most are fast release nutrient sources that can be used anytime throughout the gardening cycle giving the gardener a wide range of different combinations of major nutrients, trace minerals and growth promoting elements.

Applying Liquid Fertilizers

Unless specifically mentioned on the container most liquid fertilizers are concentrated and need to be mixed or dissolved in water before application.

Because liquid fertilizers supply plants with instant nutrients they are most useful during periods of increased need, like transplanting or when the fruit is setting. Leaf crops do well with a light application weekly or every other week.

They can be applied using a sprayer, watering can, open container or in an irrigation system.

One of the advantages of liquid fertilizers is their use as foliage feeders, however not all liquid fertilizers are suitable for this purpose and should only be used as foliage feed if the instructions on the container indicate to do so.

Apply until the liquid drips off the leaves. Concentrate the spray on leaf undersides, where leaf pores are more likely to be open. It is important to dilute liquid fertilizer to the recommended amount especially when applying to soft, sensitive and young plants to reduce the chance of fertilizer burn.

You can also water in liquid fertilizers around the root zone. A drip irrigation system can carry liquid fertilizers to your plants. The best times to spray are early morning and early evening, when the liquids will be absorbed most quickly and won't burn foliage. If possible choose a day when no rain is forecast and temperatures are not too extreme.

Below are examples of organic liquid fertilizers you can make at home:

Banana Peels

Banana peels come with loads of potassium. Fertilizers that are rich in potassium are good for flowers and fruit plants. This, however, may not be appropriate for foliage plants such as spinach and lettuce.

Use banana peel organic liquid fertilizer on squash plants and tomatoes. In a mason jar, put the banana peels, fill it with water, and cork it. Allow it to sit for three days and use it on your garden.

Eggshell Fertilizer

Eggshells are rich in calcium and contain a small percentage of potassium. Crush them, put in a mason jar, and fill it with water. Let it sit for a week and use the water on your plants. It is ideal for tomatoes and houseplants. Calcium helps prevent blossom-end rot.

Vegetable Cooking Water

After boiling vegetables, do not dispose it off. Let it cool then dilute it with water. Use it on houseplants and at the organic garden. To avoid the bad smell, use it all at once. This water is rich in vitamins that the vegetables lose during the cooking process.

Epsom Salts

Epsom salts contain sulfur and magnesium. In a gallon of water, add a tablespoon of Epsom salts. Put the mixture in a sprayer and use it on peppers, tomatoes, roses, and onions. Spray them two times a month.

Weed Tea

Unseeded weeds can be used to make fertilizer. Add grass clippings to make it richer. Grass and weeds contain nitrogen and the water accelerates its breakdown to make the nutrients available.

Place grass and weed clippings in a 5-gallon container and add water. Ensure the water goes slightly above the components, cover and let it sit for three days. Drain the liquid and dilute with water in a ratio of one to ten. Use it to water plants in the organic garden. Put the remaining components in the compost.

Droppings Tea

Collect chicken, rabbit, or goat droppings, put them in a bucket and add water until it's slightly above the droppings. Let it sit for two to three days and drain the water. Dilute it with a one to twenty water ratio and use it to water plants in the organic garden. Droppings tea is rich in nitrogen.

Compost Tea is a Liquid Gold fertilizer for flowers, vegetables and houseplants. Compost Tea, in fact, is all the rave for gardeners who repeatedly attest to higher quality vegetables, flowers, and foliage. Very simply, it is a liquid, nutritionally rich, well-balanced, organic supplement made by steeping aged compost in water. But its value is amazing, for it acts as a very mild, organic liquid fertilizer when added at any time of the year.

What is so wonderful about Compost Tea is that it can be made right at home from your own fresh, well-finished compost. The only requirement is that the compost you use is well broken-down into minute particles. This usually means that the organic materials have decomposed over a period of time so that their appearance is very dark with the texture of course crumbly cornmeal. Oh, and the fragrance is like that of rich soil in a forest.

Don't have such compost yet? Well, dig deep down inside your bin, near the bottom. This is where organic material will be most decomposed and fresh. All you need is a good shovelful for a 5-gallon bucket of Compost Tea.

Compost Tea makes you Green All Over

Leachate is actually a by-product of composting and worm composting. It is a liquid that forms in the bottom of most bins, most likely unseen by you (unless collected from a worm bin), but well-known by all the microbes and critters, including worms, who live at the bottom of your pile and in the soil. This stuff is like a fantastic smoothie or a good cup of espresso to them.

A fairly new phenomenon to gardening is the deliberate creation of Liquid Gold: Compost Tea. Researchers have determined exacting and scientific ways to brew it. The result has been the creation and promotion of Compost Tea brewing equipment, available at fine garden centers or on the internet. Some garden centers, in fact, have begun "brewing" the tea in large batches so that customers can draw-off what they need by the gallon.

The homeowner is not obligated to use exacting methods to get some very fine tea. On this website we offer a very simple, practical, and fast way to make up a batch. All you need is a couple of buckets, a shovelful of fresh finished compost, water and a straining cloth such as cheesecloth or burlap.

Good Reasons to use Compost Tea

1. Increases plant growth

It is chock full of nutrients and minerals that give greener leaves, bigger and brighter blooms, and increased size and yield of vegetables.

2. Provides nutrients to plants and soil

The fast-acting nutrients are quickly absorbed by plants through their leaves or the soil. When used as a foliar spray plant surfaces are occupied by beneficial microbes, leaving no room for pathogens to infect the plant. The plant will suffer little or no blight, mold, fungus or wilt.

3. Provides beneficial organisms

The live microbes enhance the soil and the immune system of plants. Growth of beneficial soil bacteria results in healthier, more stress-tolerant plants. The tea's chelated micronutrients are easy for plants to absorb.

4. Helps to suppress diseases

A healthy balance is created between soil and plant, increasing the ability to ward off pests, diseases, fungus and the like. Its microbial functions include: competes with disease causing microbes; degrades toxic pesticides and other chemicals; produces plant growth hormones; mineralizes a plant's available nutrients; fixes nitrogen in the plant for optimal use.

5. Replaces toxic garden chemicals

Perhaps the greatest benefit is that compost tea rids your garden of poisons that harm insects, wildlife, plants, soil and humans. It replaces chemical-based fertilizers, pesticides and fungicides. And, it will never burn a plant's leaves or roots. Finally, you save money.

6. Makes you a Green Planetary Citizen

Compost tea is just another way to feel good about respecting the earth in your own yard and garden. It allows you to be less a consumer of harmful products and more a resourceful gardener.

Tea Making Tips

The following factors will determine the quality of the finished tea:

1. Use well-aged, finished compost

    Unfinished compost may contain harmful pathogens and compost that is too old may be nutritionally deficient. Compost tea and manure tea are not the same thing. Manure teas may be made in the same way but are not generally recommended as foliar sprays and are not as nutritionally well-balanced.)

2. Using well-made, high quality compost you can brew up a mild batch in as little as an hour or let it brew for a week or more for a super concentrate.

    A good median is to let the tea brew for 24-48 hours. When it begins to smell "yeasty" you can stop and apply it to your plants.

3. Recent research indicates that using some kind of aeration and adding a sugar source (unsulphered molasses works well) results in an excellent product that extracts the maximum number of beneficial organisms. This aeration is crucial to the formation of beneficial

bacteria and the required fermentation process. For the simple bucket-brewing approach, simply stir the tea a few times during those hours or days it is brewing.

4. You can add all kinds of supplements like fish emulsion or powdered seaweed.

This turns the tea into a balanced organic fertilizer.

Use of Compost Tea

1. As A Root Drench?

Can be used unfiltered by applying directly to the soil area around a plant. The tea will seep down into the root system. Root feeding is not affected by rainy weather.

2. As A Foliar Spray?

Strain tea thru a fine mesh cloth (cheesecloth, burlap, even an old shirt). Then dilute it with de-chloronated water, if possible, or good quality well water. Use a ratio of 10 parts water to 1 part tea. The color should be that like weak tea. Add 1/8 tsp vegetable oil or mild dish-washing liquid per gallon to help it adhere to leaves.

Method of application and weather - A pump sprayer or misting bottle works better than hose-end sprayers for large areas or for foliar feeding as they don't plug up as easily. The beneficial miroorganisms are somewhat fragile so it is important to note you should avoid very high presure sprayers for appliction. Re-application after rain is necessary and one should avoid applying to the leaves during the heat of the day.

Benefits of the Organic Liquid Fertilizer

1. Economic advantage

   ○ Cost effective: Improve the economic wellbeing of growers and farmers in particular and the society in general.

2. Social advantage

   ○ No toxic effect: there is no any health hazard effect to human beings because there is no any toxicity effect.

3. Crop Productivity advantages

   ○ Controlled growth: does not over- stimulate to exceptional growth which can cause problems and require more work.

   ○ Enhance/speed up the crops maturity date: it expedites new bud initiation and also maturity of plants by up to 10 days.

   ○ Other quality improvements on the crop: it improves head size and color/texture of leaves while also increasing vase life after flowers are cut.

   ○ Stronger plants and grass: a multipurpose liquid concentrate that promotes vigorous growth, increased root development and improved stress and disease tolerance.

- For rapid uptake: Their small particle size and liquid formulations also allow for rapid uptake when applied directly to leaves (foliar feeding).

4. Soil related advantages

- Better for the soil: provides organic matter essential for microorganisms. It is one of the building blocks for fertile soil which is rich in humus.

- Nutrient release: slow and consistent at a natural rate that plants are able to use. Since microbes must break down the material no danger of over concentration of any element can be caused; and also it constitutes full of both macro and micro nutrients.

- Trace minerals: typically present in a broad range, providing more balanced nutrition to the plant.

- Open up the soil pores so that the soil microbes will multiply and begin to release nutrients, as crops use them.

- Will not burn: safe for all plants with no danger of burning due to salt concentration.

- Long lasting: doesn't leach out since the organic matter binds to the soil particles where the roots have access to it.

- Encourages soil life: Microbes convert the organic matter to the form of nutrients that plants need. Earthworms feeding on organic materials are eaten and loosen the soil.

- Specific formulas: formulated in liquid form which enables it to percolate easily to the soil and can adapt to any application by changing the ingredient blend. Pre-blended formulas or individual items allow flexibility for plant preferences or needs.

5. Environmental advantage

- Beneficial to the environment: it is free from any harmful residues or it will not cause pollution due to runoff from irrigation or rain. It is environmentally friendly and it addresses the requirements of eco-friendly.

Poultry Litter

Poultry litter is organic waste produced from chickens and turkeys like manure, spilled feed, feathers, and bedding materials.

Poultry litter is the mix of bedding material, manure and feathers that result from intensive poultry production. Broiler litter makes up the vast majority of litter produced in Australia, with an estimated 738,000 tonnes, or 1.66 million cubic metres of broiler litter produced each year, which equates to 1.72 kg of litter per broiler every seven weeks.

Litter Materials

Regional availability of dry organic materials dictates which litter material poultry growers will use. The most commonly used materials on the floor of sheds are sawdust, wood shavings, rice hulls, straw and paper products. The litter material is spread approximately 5 cm deep and can serve several flocks, although single flock clean-outs are still very common for Australian broiler sheds.

Table: Type and volume (m³) of litter material and volume of litter produced in Australia.

Litter							Total
		Shavings	Sawdust	Rice hulls	Straw	Paper	
Type	New	486,065	236,370	188,325	43,420	2,970	957,150
	Used	774,560	470,070	329,860	80,030	5,550	1,660,470

Choice of Litter Materials

Hard fibre litter materials (wood shavings) have been demonstrated to improve gizzard development and improve feed conversion efficiency, without having an effect on feed intake or weight gain. The hard fibre is thought to stimulate gut development, improve nutrient digestibility and alter the composition of the gut microflora of chickens ingesting it. While the choice of litter materials is affected by factors such as availability and cost, hard fibre materials have the twin advantages of assisting gut development and, seemingly, reducing the risk of Marek's disease.

Composition

Figure: Caked

Litter is broadly comprised of proteins, carbohydrates, lipids and fats. Carbohydrates are responsible for the majority of biodegradable materials in the form of cellulose, starch and sugars. After it has been removed from the shed, litter forms free-flowing granular material, which includes varied proportions of large, caked pieces. The chemical and physical composition of litter is highly variable due to differing bird species, diets, bedding retention times and other farm management practices.

Litter Chemistry

Two of the main elements affecting litter chemistry are Nitrogen and Phosphorous.

Nitrogen (N)

Figure: Free-flowing

Poultry manure contains two main forms of Nitrogen (N): uric acid and undigested proteins, which represent 70 and 30 percent respectively of the total nitrogen.

Under aerobic conditions, uric acid and undigested proteins break down into ammonium, which is probably why it has often been referred to as 'hot' waste. The degradation process occurs quickly, with microorganisms being a fundamental component in determining the rate of conversion. Once applied to land, ammonium quickly converts to nitrite and finally nitrate, the most readily available form of N that plants can use.

In contrast, anaerobic decomposition of uric acid and undigested proteins is slower and results in the majority of N in litter being in the ammonium form. Often the storage of litter results in 50 to 90 per cent of the total N being present as ammonium by the time it was applied to land.

Phosphorous (P)

Figure: Approximately half a shed's worth (15,000 birds) of litter

Phosphorus (P) concentrations in broiler diets are maintained to ensure rapid animal growth, and consequently manure usually has high P concentrations. Reported P concentrations in litter are variable, ranging from 9.8 to 27.1 g/kg, with the majority in a soluble form.

Table: Poultry litter analysis

Characteristic	Average	Range
pH	8.1	6.0 – 8.8
Electrical conductivity^ (dS/m)	6.8	2.0 – 9.8
Dry matter (%)	75	40 – 90
Nitrogen N (% of dry matter)	2.6	1.4 – 8.4
Phosphorus P (% of dry matter)	1.8	1.2 – 2.8
Potassium K (% of dry matter)	1.0	0.9 – 2.0
Sulphur S (% of dry matter)	0.6	0.45 – 0.75
Calcium Ca (% of dry matter)	2.5	1.7 – 3.7
Magnesium Mg (% of dry matter)	0.5	0.35 – 0.8
Sodium Na (% of dry matter)	0.3	0.25 – 0.45
Carbon C (% of dry matter)	36	28 – 40
Weight per m³ (kg)	550	500 – 650

Electrical conductivity is a measure of salinity, measured as a 1:5 suspension in water.

Like any animal manure, litter is also a potential source of pathogens and must be handled and used appropriately. There are no viral or protozoal agents present in Australian poultry that can be considered a serious or major risk to human health. This means that viral or protozoal agents are not a major human health risk in poultry litter re-use scenarios.

Current and Future Litter Reuse

Application of litter directly onto land provides a convenient mechanism for disposal and is the most commonly used waste management option. Litter acts both as a fertiliser and soil conditioner, unlike inorganic fertilisers that do not supply soil organic matter to soils. It is estimated that in excess of 90% of litter is spread on land that is close to the grower and, if used responsibly, has few environmental impacts. Currently, most Australian growers receive small profits from the sale of litter, or at least trade the litter for sheds to be cleaned and the litter taken away. For some poultry producing regions in Australia, land application of litter is becoming less cost effective, predominantly due to restrictions on land availability and the cost of transporting litter.

Litter has significant energy value, which is comparable with wood and half that of coal. As a result, power plants overseas have been developed using litter as the primary fuel for heat generation and subsequent electricity production. A host of other value adding technologies are also being developed to capture the energy and nutrients contained in litter while improving waste management for the poultry industry.

Organic Manure

Manures may be defined as materials which are organic in origin, bulky and concentrated in nature and capable of supplying plant nutrients and improving soil physical environment having no definite chemical composition with low analytical value produced from animal, plant and other organic wastes and by products.

Organic manures are included well rotten farm yard manure (FYM), compost, green manures etc. Generally farm yard manures and composts are the decomposed products of agricultural by-products (animals and crops). Whereas green manures may be defined as materials which are un-decomposed green plant tissues susceptible to decomposition in the soil after incorporation.

Classification of Organic Manures

We know that organic manures are of different types i.e. mainly bulky and concentrated in nature.

A simple classification scheme of organic manures is being presented below:

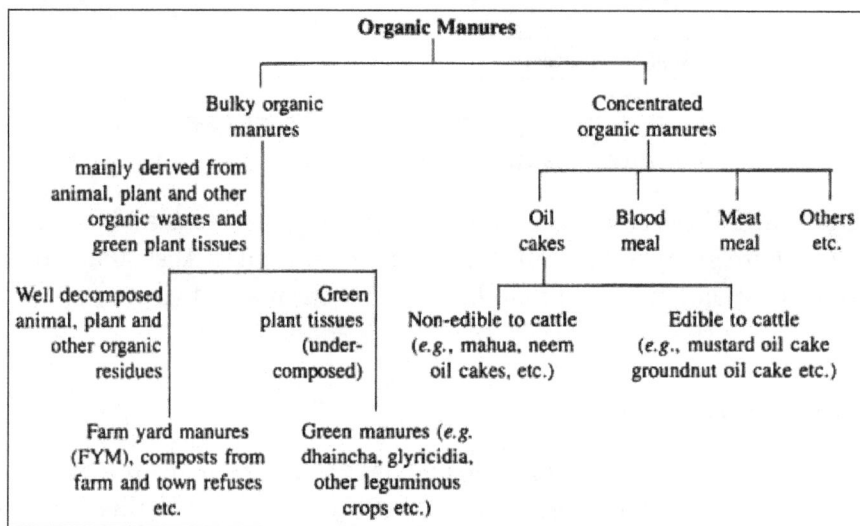

```
                            Organic Manures
               ┌──────────────────────┴──────────────────────┐
          Bulky organic                              Concentrated
            manures                                 organic manures
      mainly derived from                   ┌───────┬────────┬────────┐
      animal, plant and other              Oil     Blood    Meat    Others
      organic wastes and                   cakes    meal     meal    etc.
      green plant tissues              ┌──────┴──────┐
   ┌──────────────┬──────────┐   Non-edible to cattle    Edible to cattle
   Well decomposed      Green    (e.g., mahua, neem     (e.g., mustard oil cake
   animal, plant and   plant tissues   oil cakes, etc.)   groundnut oil cake etc.)
   other organic       (under-
   residues            composed)
       Farm yard manures    Green manures (e.g.
       (FYM), composts from  dhaincha, glyricidia,
       farm and town refuses other leguminous
       etc.                  crops etc.)
```

Bulky Organic Manures

Bulky organic manures generally contain fewer amounts of plant nutrients as compared to concentrated organic manures. The concentrated organic manures are mainly derived from raw materials of animal or plant origin.

The amount of nutrients content varies with the nature and kind of oil cakes. No definite composition of NPK and other micro-nutrients can be given. However, all oil cakes either edible or non-edible contains differential amount of N, P and K etc.

Farm Yard Manure (FYM)

The term farm yard manure refers to the well-decomposed mixture of dung, urine, farm litter (bedding material) and left over or used up materials from roughages or fodder fed to the cattle. The FYM collected daily from the cattle shed consisting of raw dung and part of the urine absorbed in the refuse.

Newly collected and stored FYM is fresh as against well decomposed FYM which has been stored for a sufficient period of time to allow its decomposition to completion.

Farm yard manure consists of two components solid phase, dung and liquid phase, urine. On an average, the animals give out three parts by weight of dung and one part by weight of urine. However, this ratio of dung and urine varies with the kind of animals.

Horses, cows and bullocks give out more dung and less urine than that of sheep, goats and pigs. An average nutrient and moisture content of different kinds of animals is given in table.

Table: Average amount of moisture and nutrient content of Different Animals

Animals	Dung : Urine Moisture		'Nutrients' (kg/t)		
	Ratio	(%)	Nitrogen (N)	Phosphorus (P_2O_5)	Potasium (K_2O)
Dairy Cattle	80 : 20	85	4.53	1.22	3.40
Feeder Cattle	80 : 20	85	5.40	2.13	3.22
Poultry	100 : 0	62	13.54	6.48	3.17
Swine	60 : 40	85	5.84	3.22	4.94
Sheep	67 : 33	66	10.42	3.17	9.83
Horse	80 : 20	66	66.75	2.03	5.98

Average Amount of Moisture and Nutrient Content of different Animals.

Urine of all animals contains more amounts of N and P as compared to dung. The urine of cows, bullocks and horses contain practically nil or traces of P. Similarly, the dung of all animals except pig is low in P.

Green Manuring

Green manuring can be defined as a practice of ploughing or turning into the soil un-decomposed green plant tissues for the purpose of improving soil physical chemical and biological environments.

Kinds of Green Manuring

The practice of green manuring is performed in different ways according to suitable soil and climatic conditions of particular area. Broadly the practice of green manuring can be divided into two types—Green manuring in situ and Green manuring by collecting green leaves and tender twigs from some other places.

Green Manuring in Situ

It can be defined as a system by which green manure crops are grown and incorporated into the soil of the same field that is to be green manured, either as a pure crop or an intercrop with the main crop. Common green manure crops in this system sun hemp (crotolariajuncea) dhaincha (sesbaniaaculeata and sesbaniarostrata), guar (cyamopsistetragonoloba), etc.

Green Manuring through Collection of green Plant Tissues from other Places

It refers to turning into the soil green leaves and tender green twigs collected from outside the field to be green manured. The common green manure crops, are Glyricidia (Glyricidiamaculata), Karanja (Pongamiapinnata) etc.

Advantages of Green Manuring

There are various advantages of green manuring in relation to soil fertility which are as follows:

1. It increases the organic matter regime of the soil and there by modifies soil physical, chemical and biological environments. In fact, this stimulates the activity of soil micro-organisms.

2. The green manure crops help for returning the different plant nutrients to the surface soil layer from the sub-surface soil layer.

3. It improves the soil structure, aeration status, permeability and infiltration capacity of soil.

4. It reduces the soil loss caused by run-off and erosion.

5. Due to green manuring the nutrient regimes can be improved and restored otherwise be lost by leaching.

6. Green manure crops have some residual effect in relation to supply of different plant nutrient and thereby it helps for the better growth to the next crop.

Disadvantages of Green Manuring

The adoption of green manuring in a improper way and application or incorporation of green manure crops without proper soil and water management leads to the following deleterious effects:

1. Under rainfed conditions where rainfall is limiting, the proper decomposition of the green manure crops may not take place and thereby benefits of green manuring may not be achieved satisfactorily.

2. Sometimes the cost of green manuring crops may be more than that of chemical nitrogenous fertilizers and in that situation green manuring may not be economical.

3. There is a change of occurring diseases and insects in the field crops.

4. Due to decomposition of green manure crops, various toxic substances like organic acids e.g. butyric acid, propionic acid etc. and toxic gases like methane (CH_4) and others etc. are liberated which affects the root growth of growing plants and thereby affect the growth and yield through inhibition of nutrient absorption by the plant.

Concentrated Organic Manures

Concentrated organic manure may be defined as a material of organic origin derived from raw materials of animal or plant, without bulky in nature having no definite composition of plant nutrients.

Some most common such organic manures are oil cakes edible to cattle (e.g. mustard oil cake, groundnut oil cake, till oil cake etc.) and non-edible to cattle (e.g. neem oil cake, mahua oil cakes etc.); blood-meal, fish manure, bone meal etc.

Since the sources of availability of concentrated organic manures are different and hence they do not contain definite amount of nutrient elements. However, concentrated organic manures are easy to handle and have relatively higher plant nutritive value as compared to bulky organic manures. Besides these, it is quick-acting organic manure when incorporated into the soil.

Characteristics of Organic Manures

All these manures are bulky in nature as well as concentrated nature and supply:

1. Plant nutrients in small quantities and

2. Organic matter in large amounts.

Since it contains two components (plant nutrients and organic matter), when it is applied into the soil it will act as follows:

1. Organic manures supply primary, secondary and micro-nutrients to plants which are liberated in an available forms during the process of mineralization carried out by different micro-organisms.

2. Organic manures also supply organic matter to the soil and hence improve the physical condition of the soil like soil structure, aeration, water holding capacity etc.

3. It also stimulates the activity of different soil micro-organisms through the supply of energy.

4. It improves the buffering and exchange capacities of soil and also influences the solubility of soil minerals as well as mineral nutrients in soil.

5. It also forms chelates which also help for the nutrition of plants.

6. It also regulates the thermal regimes of the soil.

Factors Affecting the Composition of Organic Manures

There are various factors which can affect the composition as follows:

1. Origin of Manure: Sheep and poultry manures are somewhat richer in plant nutrients than cow, horse and pig manures.

2. Types of Food Consumed by Animals: This is one of the most important factors that determine the manure quality. As for example, the richer the food in proteins, the richer will be the manure in nitrogen.

3. Age and Condition of Animals: The manure of young animals is not so rich like that of matured animals because young animals retain more nutrients for their growth than that of old or matured ones.

4. Species of Animals: The composition of nutrient contents varies with the ruminant and non-ruminant animals.

5. Nature and Amount of Litter: The composition of FYM varies with the nature and amount of litter used for animals (e.g. paddy straws, wheat straws etc.)

6. Function of the Animal: Animals producing milk and wool absorb large amount of nutrients from their food than that of working draft animals. Therefore, manure from bullocks generally contains more nutrients as compared to milch cows.

7. Handling and Storage of Manures: Loss of potash occurs if any drainage is allowed to escape from the manure heap. Therefore, improper handling and storage leads to losses of plant nutrients from the manures.

Reactions of Organic Manures in Soils

It is found that both bulky and concentrated organic manures contain some amount of plant nutrients including macro- and micro-nutrients, of which organic nitrogen content is likely to be dominant. The organic forms of soil nitrogen occur as consolidated amino acids, proteins, amino sugars etc.

When Organic manures like FYM, composts, oil cakes, green manures etc. are added to the soil, the microbial attack to these materials takes place and results complete disappearance of the organic protein with the remainder of the nitrogen being changed into inorganic form of nitrogen through the process of mineralization.

$$\text{Proteins} \rightarrow R - NH_2 + CO_2 + \text{energy} + \text{other products}$$

(presentin

organic manure)

$$P - NH_2 + HOH \rightarrow \underset{\searrow + H_2O}{NH_3} + R - OH + \text{energy}$$

$$NH_4^+ + OH^-$$

(release of ammonium in the soil)

The released ammonium NH_4^+ is subject to following changes:

1. It may be converted to nitrites and nitrates through the process of nitrification carried out by micro-organisms,

$$2NH_4^+ + 3O_2 \xrightarrow[\text{oxidation}]{\text{enzymic}} 2NO_2^- + 2H_2O + 4H^+ + \text{energy}$$
$$2NO_2^- + O_2 \xrightarrow[\text{oxidation}]{\text{enzymic}} 2NO_3^- + \text{energy}$$

2. It may be absorbed directly by the plants.

3. It may be utilized by hetero-trophic organisms in further decomposing organic carbon residues.

4. It may be fixed in the lattice of certain expanding-type clay minerals.

5. It could be slowly released back to the atmosphere as elemental nitrogen.

Mineralization and immobilization of nitrogen or any other nutrient elements occur continuously in microbial metabolism and the magnitude and direction of the net effect are greatly influenced by the nature and amount of organic manures added.

Normally organic manures are applied to the soil as a source of fertilizer nitrogen should contain about 1.5 to 2.0 percent of the dry weight of the manures in order to meet the needs of the soil microorganisms, otherwise little or no nitrogen will be released for the use of plants.

The carbon nitrogen ratio (C: N ratio) in the organic manures remaining in the soil after consuming by the soil micro-organisms is approximately 10:1.

Therefore, different organic manures containing variety of organically bound nutrients like P, S and other micro-nutrients etc. are subject to transformation in soils similar to that of mineralization and immobilization processes of nitrogen and releases inorganic forms of nutrients in soils which become available to plants.

Manure Management in Organic Systems

Livestock manure is a key fertilizer in organic and sustainable soil management. Manure provides plant nutrients and can be an excellent soil conditioner. Properly managed manure applications recycle nutrients to crops, improve soil quality, and protect water quality. It is most effectively used in combination with crop rotation, cover cropping, green manuring, liming, and the addition of other natural or biologically-friendly fertilizers and amendments.

Use of manure imported from conventional farming operations is allowed by National Organic Program (NOP) standards. There are, however, application restrictions. Manure may only be used in conjunction with other soil-building practices and be stored in a way that prevents contamination of surface or ground water. Many certifiers specify that manure application must not exceed "agronomic application rates", which means the amount applied must be less than or equal to the requirements of the crop. Manure cannot be applied when the ground is frozen, snow-covered, or saturated.

The NOP regulation specifies that "raw" fresh, aerated, anaerobic, or "sheet composted" manures may only be applied on perennials or crops not for human consumption, or such uncomposted manures must be incorporated at least four months (120 days) before harvest of a crop for human consumption, if the crop contacts the soil or soil particles (especially important for nitrate accumulators, such as spinach). If the crop for human consumption does not contact the soil or soil particles (e.g. sweet corn), raw manure can be incorporated up to 90 days prior to harvest. Biosolids, sewage sludge, and other human wastes are prohibited. Septic wastes are prohibited, as well as anything containing human waste.

Composted plant and animal manures are those that are produced by a process that:

1. Established an initial C:N ratio of between 25:1 and 40:1; and

2. Maintained a temperature of 131°F to 170°F for 3 days using an in-vessel or static aerated pile system; or

3. A temperature of between 131°F and 170°F for 15 days using a windrow composting system, during which period, the materials must be turned a minimum of five times.

Heat-treated, processed manure may be used as a supplement to a soil-building program, without a specific interval between application and harvest. Producers are expected to comply with all applicable requirements of the NOP regulation with respect to soil quality, including ensuring the soil is enhanced and maintained through proper stewardship.

According to the NOP's July 17, 2007 ruling, "processed manure products must be treated so that all portions of the product, without causing combustion, reach a minimum temperature of either 150 °F (66 °C) for at least one hour or 165 °F (74 °C), and are dried to a maximum moisture level of 12%; or an equivalent heating and drying process could be used." To achieve equivalency status, processed manure products can not contain more than 1×10^3 (1,000) MPN (Most Probable Number) fecal coliform per gram of processed material sampled and not contain more than 3 MPN Salmonella per 4 gram sample of processed manure.

As always, organic vegetable growers should get label information and check with their certifiers before using purchased compost or processed manure products.

Some manures are contaminated with hormones, antibiotics, pesticides, disease organisms, heavy metals, and other undesirable substances. Many of the organic compounds, pathogens, protozoa, or viruses can be eliminated through high-temperature aerobic composting. Caution is advised, however, as some disease causing agents, e.g. Salmonella and E. coli bacteria, may survive the composting process. Manure and compost testing is available through commercial labs and is recomended in situations where there is any doubt about the purity of manures. Manure testing is required by the European Union and Canadian standards. The possibility of transmitting human diseases discourages the use of fresh manures and even some composts as pre-plant or sidedress fertilizers on vegetable crops. Apply animal manures at least 90 or 120 days, as applicable, prior to harvest of any crop that could be eaten without cooking.

Best management practices recommended for manure are as follows:

1. Avoid manuring after planting a crop to be harvested.

2. Incorporation before planting is recommended.

3. Do not use dog or cat (fresh or composted) because these species share many parasites with humans.

4. Wash all produce from manured fields thoroughly before use.

Cautions or concerns include the following:

1. Manures imported from conventional farms can contain residues from hormones or pesticides. In rare cases, carryover of persistent herbicides can occur. Most herbicides break down rapidly after application or during normal composting. However, some of those in the pyridine carboxylic acid group such as clopyralid, which is commonly used on grass lawns, break down slowly, even during composting, and are not degraded when ingested by animals because they pass into the urine quickly. Application of manures or composts

derived from grass treated with clopyralid is restricted during the "growing season of application" for all farms, not just those that are organic.

2. Heavy metals (e.g., As, Cu, and Zn) are fed to livestock and then added to soils in the form of manures. Unlike sludge, metal content does not influence manure application rates to soils but should be considered as metals persist in the soil and will accumulate with repeat application. Concerns over heavy metals, other chemical contaminants, and salinity are most often raised in association with poultry litter. Under federal organic standards, certifiers may require testing of manure or compost if there is reason to suspect high levels of contamination.

3. Weed seeds and plant diseases can be effectively controlled by high temperature aerobic composting of manures.

Manure Handling: Raw Stacked or Composted

The NOP regulation also requires that manure and other fertility inputs must be managed so that they do not contribute to contamination of crops, soil, or water by excess nutrients, pathogens, heavy metals, or residues of prohibited substances. Whether animals are raised on farm or manures are imported, organic farmers are likely to need to store manure on farm prior to application. Proper manure storage conserves nutrients and protects surface and groundwater. Storing manure can be as elaborate as keeping it under cover in a building, or as simple as covering the manure pile with a tarp. The important point is keeping the pile covered and away from drainage areas and standing water. The storage location should also be convenient to your animals and crop production.

When you are looking for organic forms of nutrients for crop production, manure and manure composts are two of the logical choices. Composting is more than just piling the material and letting it sit. Composting is the active management of manure and bedding to aid the decomposition of organic materials by microorganisms under controlled conditions. Weed and disease problems associated with raw manures can be alleviated with proper composting. Use of composted manures can also reduce P transport to water bodies.

Organic producers making their own compost must keep records of their composting operation to demonstrate that the compost was produced according to the definition cited above. If the compost is purchased, the grower should ask for documentation from the supplier showing that the compost meets NOP requirements. Keep this documentation, along with purchase receipts, with your other records. If the compost is 100% plant-based, without any animal excrement or by-products, there is no requirement for heating or turning.

Table: Comparison of composted and raw manures.

Compost	Manure
Slow release form of nutrients	Usually higher nutrient content
Easier to spread	Sometimes difficult to spread
Lower potential to degrade water quality	Higher potential to degrade water quality

Less likely to contain weed seeds	More likely to contain weed seeds
Reduced pathogen levels (e.G. Salmonella, e. Coli)	Potential for higher pathogen levels
Higher investment of time or money	Lower investment of time or money
More expensive to purchase	Less expensive to purchase
Fewer odors (although poor composting conditions create foul odors)	Odors sometimes a problem
Improves soil tilth	Improves soil tilth

Managing Nutrients in Manure

Manure nutrient contents are highly variable and growers must be able to understand and reduce this variability to make the best agronomic and environmental use of these resources. Manure must be carefully managed to prevent over-or under-application and to account for the cumulative environmental effects of application as well as storage. Balancing crop nutritional needs with manures is an ongoing challenge. Finding out about manure composition is critical to its efficient use. Applying too little can lead to inadequate crop growth because of lack of nutrients. Over-application can reduce crop quality and increase the risk of plant diseases. Over-application will also increases the risk of contaminating surface or groundwater.

There are three main sources of variability and uncertainty when using manure:

1. Nutrient and moisture content of the manure.

2. Material heterogeneity and application variability.

3. Availability of nutrients to crops.

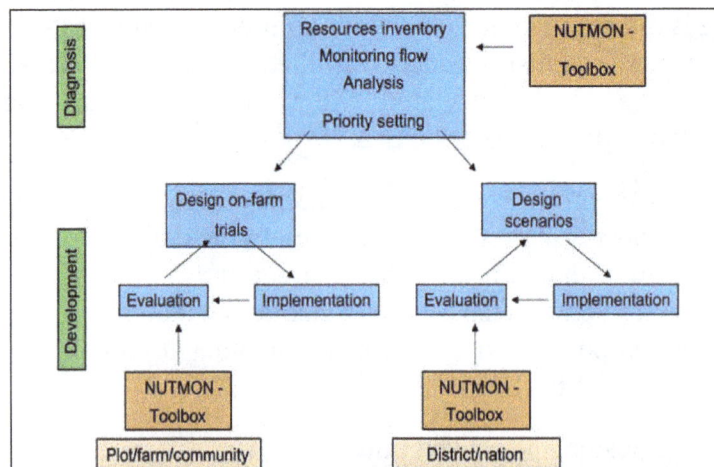

Nutrient flow from manure resources to storage facilities and then to field. Nutrients
can be lost from all locations but only those arriving on the field have the chance to feed plants.

Manure composition varies with the species of animal, feed, bedding, and manure storage practices. Table shows typical published values for livestock manure. These values may not accurately represent your situation. Nutrient values can vary by a factor of two or more from the values listed in table. This is why it is important to test materials applied instead of guessing.

Table: Typical nutrient content of manure. Because of variability between farms, individual manure analysis is preferable to the estimates below.

	% Dry Matter	Ammonium–N	Organic–N	P_2O_5	K_2O
Slurry Manure	(lb. of nutrient per 1,000 gallons of manure)				
Dairy	8	12	13	25	40
Beef	29	5	9	9	13
Swine (finisher, wet-dry feeder)	9	42	17	40	24
Swine (slurry storage, dry feeder)	6	28	11	34	24
Swine (flush building)	2	12	5	13	17
Layer	11	37	20	51	33
Dairy (lagoon sludge)*	10	4	17	20	16
Swine (lagoon sludge)	10	6	16	48	7
Solid Manure	(lb. of nutrient per ton of manure)				
Beef (dirt lot)	67	2	22	23	30
Beef (paved lot)*	29	5	9	9	13
Swine (hood barns)	57	4	13	20	
Dairy (scraped earthen lots)	46	3	14	11	16
Broiler (litter from house)	70	15	60	27	33
Layer	40	18	19	55	31
Turkey (grower house litter)	70			15	30
Liquid Effluent from lagoon or holding pond	(lbs. of nutrient per acre-inch)				
Beef (runoff holding pond)	0.25	71	8	47	92
Swine (lagoon)	0.40	91	45	104	189
Dairy (lagoon)	2	317	362	674	1082

Value based upon ASAE, 2005, D384.2; Manure Production and Characteristics with exception of those marked with an.

Manure Sampling and Testing

Commercial laboratories can measure the nutrients in manure and save you from guessing based on table values. Testing laboratories typically charge from $30 to $60. It is important to use a laboratory that routinely tests animal manure, as they will know the correct type of analysis to use. Extension offices can provide you with publications that list manure testing laboratories in most regions.

A nutrient analysis is only as good as the sample you take. The best time to sample by far is right before you apply the material because N loss in storage is accounted for. Also, if you use manure repeatedly from the same source, you can develop a running average analysis of that manure (over a 3+ year period). A running average is more likely to be accurate than a single sample taken from a storage pile or lagoon. Samples must be fresh and representative of the manure. Follow these steps:

1. Ask the laboratory what type of containers they prefer and make sure the laboratory knows when your sample is coming. Laboratories should receive samples within 48 hours of collection. Plan to collect and send your sample early in the week so the sample does not arrive at the lab on a Friday or a weekend.

2. If you have a bucket loader and a large amount of manure, use the loader to mix the manure before sampling.

3. Take 10–20 small samples from different parts and depths of the manure pile to form a composite sample. The composite sample should be about 5 gallons. The more heterogeneous your pile, the more samples you should take.

4. With a shovel or your hands thoroughly mix the composite sample. You may need to use your hands to ensure complete mixing. Wear rubber gloves when mixing manure samples with your hands.

5. Collect about one quart of manure from the composite sample and place in an appropriate container.

6. Freeze the sample if you are mailing it. Use rapid delivery to ensure that it arrives at the laboratory within 24–48 hours. You can refrigerate the sample if you are delivering it directly to the lab.

Sample Information		
Sample Name: M 4419		**Storage System:** Solid
Material: Dairy		**Type of Storage:** stacked pile-outside
Treatment: None		**Type of Bedding:** sawdust/shavings/bark

Laboratory Analysis

Moisture: 67.20%

Dry Matter: 32.80%

	Total Nutrients lbs/ton	Estimated Available Nutrient Credits for Manure:		
		In 1st Year of Application lbs/ton	If Applied 2 Consecutive Yrs lbs/ton	If Applied 3 Consecutive Yrs lbs/ton
Total Nitrogen (Injected)	8.53	3.41	4.27	4.69
Total Nitrogen (Surface Applied)	8.53	2.56	3.41	3.84
Total Phosphorus as P_2O_5	3.30	1.98	2.31	2.48
Total Potassium as K_2O	4.88	3.90	4.39	4.64
Sulfur	2.10	1.26	1.47	1.57
Estimated Value of Available Nutrients in Surface Applied Manure[1]		$7.85	$9.43	$10.23

Additional Tests

NH$_4$-N: 1.97 lbs/ton

Ash: 64.31%

Additional Information

1   Value based on commercial fertilizer costs as of 8/6/2008:
     N (urea) $0.88/lb
     P_2O_5 (Triple Superphosphate) $1.00/lb
     K_2O (Potash) $0.68/lb
     S (Elemental Sulfur) $0.77/lb

Figure: Example of a manure analysis report. Note, the report includes "additional information" about the relative value of nutrients which is subject to change. By convention, available nutrient contents are expressed in terms of reference materials used in fertilizer labels

Laboratories report results on an as-received or a dry weight basis. As-received results usually are reported in units of lb/ton, while dry weight results usually are reported in percent, ppm, or mg/kg. The "as-received" results, as shown above, are easily used to determining application rates. Dry-weight results can be used to compare analyses over time and from different manure sources.

To convert manure analyses reported on a dry-weight basis (in percent) to an as-received basis (in lb/wet ton), multiply by 20 to convert the dry weight percent to lb/ton; then multiply by the decimal equivalent (23%/100) of the solids content.

Example: For beef manure at 23% solids and 2.4% nitrogen (N) on a dry weight basis:

- Step 1. 2.4% × 20 = 48 lb N/ton dry weight.

- Step 2. 48 lb N/ton dry weight × 0.23 = 11 lb N/ton as-is.

Analyses typically include total nitrogen, ammonium nitrogen (NH_4^+-N), total phosphorus, total potassium, electrical conductivity, and solids (dry matter). If the manure is old or has been composted you may also want to test for nitrate–N. Total carbon (C) and pH are also useful measurements. Total C can be used to determine the C:N ratio and predict whether or not manure addition is likely to cause nitrogen immobilization. Manure with a C:N ratio greater than 25 is likely to 'tie up' or immobilize nitrogen when you apply it to the soil and stimulate a flush of growth by bacteria and fungi. Bedded manures typically have higher C:N ratios.

Manure Application Rates

When application rates of manure are based on providing adequate nitrogen for crop growth, added phosphorus and potassium levels will often exceed crop need, so manure should not ve the sole N source in an organic system. Excess levels of soil P can increase the amount of P in runoff, increasing the risk of surface water pollution. Many crops can handle high levels of K, but livestock can be harmed by nutrient imbalances if they consume a diet of forages with high K levels. Annual P-based manure or compost application is the most effective method of application when soil P buildup is a concern. Phosphorus-based application rates improve water quality, but reduce the amount of manure applied per area and so increase the land base needed for manure application. Where P buildup is a concern, legumes should be included in the rotation to provide additional nitrogen.

Typically manure is applied before the most N-demanding crop in the rotation and the amount of N likely to be plant available during the year of application is estimated. Nitrogen availability from manure varies greatly, depending on the type of animal, type and amount of bedding, and age and storage of manure. Manure contains nitrogen in the organic and ammonium forms. The organic form releases N slowly, while ammonium–N is immediately available for crop growth.

Table: Manure N availability in the first year after application.

Manure type	Total N content (%)	% available N
Broiler litter	4–6	40–70
Laying hen	4–6	40–70
Sheep	2.5–4	25–50
Rabbit	2.5–3.5	20–40
Beef	2–3	20–40
Dry Stack	1.2–2.5	20–40
Separated Solids	1–2	0–20
Horse	0.8–1.6	0–20

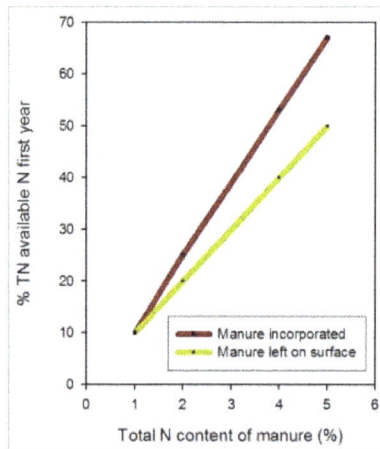

Figure: This graph can be used to predict N release based on total N content
during the first year after manure application

Solid manures contain most of their nitrogen in the organic form, but poultry manure contains substantial ammonium–N and so should not be surface applied to avoid loss of ammonia gas. Poultry and other manures that contain a large proportion of ammonium–N should be tilled into the soil the same day they are spread. Ammonia loss is greater in warm, dry, and breezy conditions where soil pH is high and is reduced in cool, wet weather. The N availability numbers in table are approximate ranges for each type of manure. Use the lower part of the ranges if ammonia losses are likely, the manure contains large amounts of bedding, or if the measured N content is lower than typical values. Use the upper range if the manure contains little bedding or if the measured N content is high. Expect first year N tie-up from manures containing less than 1% N. Horse manure or other manures with lots of woody bedding may temporarily tie up nitrogen rather than supply nitrogen for crop growth because the wood is still decaying and bacteria that break down the carbon in the wood consume nitrogen. Composting generally reduces the rate of release of manure N by as much as 50% by converting N into more biologically resistant forms.

Calculators and manure test information are a far better way to calculate application rates than using tables to estimate manure nutrient content and availability.

Monitoring Soil Nutrient Levels

Repeated applications of manure can increase the pool of slow-release nutrients and so the amount of manure needed to meet crop needs will decline over time. Farmers need to reduce the manure application rate for fields that receive repeated manure applications.

Table: Effect of 11 years of annual manure additions on the properties of a heavy clay soil planted to continuous corn silage in Vermont.

Original Level	Application Rate (tons/acre/year)				
	0	10	20	30	
Soil organic matter (%)	5.2	4.3	4.8	5.2	5.5
CEC (meq/100g)	17.8	15.8	17	17.8	18.9
pH	6.4	6.0	6.2	6.3	6.4

P (ppm)	4	6.0	7.0	14	17
K (ppm)	129	121	159	191	232
Total Pore Space (%)	n.d.	44	45	47	50

Table shows trends where repeat application of manure to a clay soil at three rates has influenced soil properties. Organic matter levels were only maintained where rates equaled 20 tons or more. At these rates P and K levels were in excess. To avoid manure-induced imbalances, continually monitor soil fertility, using appropriate soil tests. Use cover crops and lime or other supplementary fertilizers and amendments to ensure soil balance or restrict application levels if needed.

You can use basic soil tests to evaluate the soil for sufficiency or excess of other nutrients. A basic soil test includes P, K, calcium (Ca), magnesium (Mg), boron (B), pH, EC, and a lime recommendation. If you have consistently low levels of P and K and reduced crop growth, you can probably increase your manure application rates. If you have excessive levels of P and K, you should decrease or eliminate manure applications.

Soil tests can be timed to evaluate different aspects of nutrient supply. Testing the year after manures are applied to assess increased P and K supply is recommended. Table shows how test levels rise for several years due to the slow release nature of P contained in many manures.

Table: The amount of available phosphorus, expressed as fertilizer equivalent P per tonne or m_3, made available in years following manure application increases over time and vaires with the type of manure added.

	Years after application	Fertilizer equivalent P per tonne			
		n	Median	Mean	P
Cattle FYM	1	39	0.48	0.62	
	2	55	0.75	1.21	0.006
	3	23	1.1	1.82	0.018
	4+	8	2.2	2.61	0.12
Pig manure	1	11	0.44	0.56	
	2	11	0.93	0.89	0.04
Poultry manure	1	5	1.21	2.44	
	2, 3	7	5.23	5.89	0.04
Cattle slurry	1	2	-0.37	-0.37	
	2, 3	5	1.63	2.92	0.05
Pig slurry	1	3	0.82	0.87	
	2	6	0.97	1.24	0.46
	3	3	1.53	1.13	-0.84

P-values are for the mean FEP being the same as at Ct - 1; italics if $P < 0.05$. Standard errors for each mean are about 0.5 kg/t for FYM and raw cake, 0.2 for pig manure. The local digested cakes contained about 5 kg total P per tonne and raw cake 2.5 kg total P per tonne.

Late season sampling (0–12 inches or more) can be used to determine whether there is surplus nitrate-N remaining in the soil in the fall. If you apply too much manure, unused nitrate N will accumulate. When the fall and winter rains come, the nitrate will leach from the soil and become a potential contaminant in groundwater or surface water. Excess N can also harm some crops, delaying fruiting and increasing the risk of disease damage, freeze damage, and wind damage. Take a "report card" sample as you would any other soil sample, collecting soil cores at multiple spots in the field, and combining the cores together into a composite sample. If late season nitrate–N results are greater than 15–20 mg/kg, you are supplying more N than your crop needs and should reduce or avoid manure application. Report-card nitrate-N levels greater than 30 mg/kg are excessive.

Nutrient requirements for specific crops can be found in Cooperative Extension production guides or from soil test recommendations. Crop performance can be used to help fine tune application rates.

Neem: The Panacea for Organic Farming

Neem is a stout tree with a rather short stem. They are native to southeastern countries. A Neem tree generally grows up to 12-15 meters. They are best known for their medicinal properties. Neem works on pest's hormonal system and does not lead to resistance like that of pesticides. Neem produce organic acids when mixed with soil, hence can be a great choice for reducing soil pH.

Make Neem Cake

People collect neem seeds from the trees. The neem seed kernels are then crushed and neem oil is extracted.

The de-oiled residue of these seeds is called neem cake. They are high in NPK and also can be used to kill harmful nematodes. Please remember that all nematodes are not harmful to use this strategy wisely.

Neem Cake as Organic Fertilizer

Since very early neem are being used as a natural fertilizer. It had become very popular due to its dual impact of soil enhancer as well as pest-repellent. The cakes, as well as neem leaves, are used to fertilize the soil and to improve soil quality.

The followings are the nutrient content of neem seed cake:

(N)	Nitrogen	(2.0% to 5.0%)
(P)	Phosphorus	(0.5% to 1.0%)
(K)	Potassium	(1.0% to 2.0%)
(Ca)	Calcium	(0.5% to 3.0%)
(Zn)	Zinc	(15 ppm to 60 ppm)

(Cu)	Copper	(4 ppm to 20 ppm)
(S)	Sulphur	(0.2% to 3.0%)
(Mg)	Magnesium	(0.3% to 1.0%)
(Fe)	Iron	(500 ppm to 1200 ppm)
(Mn)	Manganese	(20 ppm to 60 ppm)

Neem and Soil pH

Neem can be great if you are considering decreasing your soil pH. When mixed with soil it produces organic acids. So it reduces the alkalinity of the soil. it is extensively used in cash crops like turmeric, sugarcane, banana etc. In growing flowers and vegetables, neem oil cakes are a good alternative to chemical fertilizers.

Neem oil cakes improve the appearances of fruits and vegetables. It also strengthens roots, and grow the foliage. It works very well when mixed with compost or other organic fertilizers. Applying neem with nitrogenous fertilizers can slow down the conversion process. This increases the efficiency of the soil.

Neem Cakes and Nematodes

Neem seed cakes are very useful in controlling nematode population. It also effective against many soil pathogens. Neem seeds and cakes contain nortriterpenoids and isoprenoids, which are nematicidal in nature.

Neem Cakes and Urea

Neem oil cakes have many different nutrients (NPK and micronutrients) and produce a better yield than urea.

They block soil bacteria from releasing nitrogen gas. Thus it extends the availability of soil nitrates to both short duration and long duration crops.

You can also mix neem seed cakes with urea or any other fertilizers (which have nitrogen). It will increase their efficiency by reducing their nitrogen releasing rate. So the fertilizers remain in the soil for much longer.

Different Types of Neem Cakes

Neem cakes can be found in many forms. Here are some of the most common types.

- Neem cake fertilizer.
- Neem cake powder.
- Neem cake granules.
- Neem cake manures.

Pros and Cons of using Neem Cakes

Pros of using Neem Cakes

- It is an excellent Organic Soil Amendment. It enriches the soil and can be used as other soil conditioners.

- Neem cakes are organic and natural substances. They are bio-degradable.

- Totally chemical free.

- You can mix them with other fertilizers that are organic. Neem cake is a natural nitrification inhibitor. It increases the availability of nitrogen. They improve the texture of your garden soil.

- Their organic nature and water holding capacity also help to keep the soil aerated. This is a must for better root development.

- As neem cakes are sustainable, in the long-term it is very cost efficient.

- Neem oil cakes can also work as a pest repellant. These dual effects of fertilizing and pest protection can result in amazing yields.

Cons of using Neem Cakes

Just before you jump to the conclusion, here are a few points you should consider:

- Use caution while applying neem cakes in potting mixes. Don't use more than 1%. It can cause a lack of seed germination or stunt young plants.

- Botanical insecticides, such as neem are allowed in organic production. But like any other, you should use them only as a last option. Although this is natural, it sometimes may harm some beneficial insects.

Composting

Composting is nature's way of recycling. Composting biodegrades organic waste. i.e. food waste, manure, leaves, grass trimmings, paper, wood, feathers, crop residue etc., and turns it into a valuable organic fertilizer.

Composting is a natural biological process, carried out under controlled aerobic conditions (requires oxygen). In this process, various microorganisms, including bacteria and fungi, break down organic matter into simpler substances. The effectiveness of the composting process is dependent upon the environmental conditions present within the composting system i.e. oxygen, temperature, moisture, material disturbance, organic matter and the size and activity of microbial populations.

Composting is not a mysterious or complicated process. Natural recycling (composting) occurs on a continuous basis in the natural environment. Organic matter is metabolized by microorganisms and consumed by invertebrates. The resulting nutrients are returned to the soil to support plant growth.

Composting is relatively simple to manage and can be carried out on a wide range of scales in almost any indoor or outdoor environment and in almost any geographic location. It has the potential to manage most of the organic material in the waste stream including restaurant waste, leaves and yard wastes, farm waste, animal manure, animal carcasses, paper products, sewage sludge, wood etc. and can be easily incorporated into any waste management plan.

Since approximately 45 - 55% of the waste stream is organic matter, composting can play a significant role in diverting waste from landfills thereby conserving landfill space and reducing the production of leachate and methane gas. In addition, an effective composting program can produce a high quality soil amendment with a variety of end uses.

The essential elements required by the composting microorganisms are carbon, nitrogen, oxygen and moisture. If any of these elements are lacking, or if they are not provided in the proper proportion, the microorganisms will not flourish and will not provide adequate heat. A composting process that operates at optimum performance will convert organic matter into stable compost that is odor and pathogen free, and a poor breeding substrate for flies and other insects. In addition, it will significantly reduce the volume and weight of organic waste as the composting process converts much of the biodegradable component to gaseous carbon dioxide.

The composting process is carried out by three classes of microbes:

- Psychrophiles - low temperature microbes.

- Mesophiles -medium temperature microbes.

- Thermophiles - high temperature microbes.

Generally, composting begins at mesophilic temperatures and progresses into the thermophilic range. In later stages other organisms including Actinomycetes, Centipedes, Millipedes, Fungi, Sowbugs, Spiders and Earthworms assist in the process.

Temperature

Temperature is directly proportional to the biological activity within the composting system. As the metabolic rate of the microbes accelerates, the temperature within the system increases. Conversely, as the metabolic rate of the microbes decreases, the system temperature decreases. Maintaining a temperature of 130 °F or more for 3 to 4 days favors the destruction of weed seeds, fly larvae and plant pathogens.

At a temperature of 155 degrees F, organic matter will decompose about twice as fast as at 130 degrees F. Temperatures above 155 degrees F may result in the destruction of certain microbe populations. In this case temperature may rapidly decline. Temperature will slowly rise again as the microbe population regenerates.

Moisture content, oxygen availability, and microbial activity all influence temperature. When the pile temperature is increasing, it is operating at optimum performance and should be left alone. As the temperature peaks, and begins to decrease, the pile should be turned to incorporate oxygen into the compost. Subsequently, the pile should respond to the turning and incorporation of oxygen, and temperature should again cycle upwards. The turning process should be continued until the pile fails to re-heat. This indicates that the compost material is biologically stable.

Composting microorganisms thrive in moist conditions. For optimum performance, moisture content within the composting environment should be maintained at 45 percent. Too much water can cause the compost pile to go anaerobic and emit obnoxious odors. Too little will prevent the microorganisms from propagating.

Particle Size

The ideal particle size is around 2 to 3 inches. In some cases, such as in the composting of grass clippings, the raw material may be too dense to permit adequate air flow or may be too moist. A common solution to this problem is to add a bulking agent (straw, dry leaves, paper, cardboard) to allow for proper air flow. Mixing materials of different sizes and textures also helps aerate the compost pile.

Turning

During the composting process oxygen is used up quickly by the microbes as they metabolize the organic matter. As the oxygen becomes depleted the composting process slows and temperatures decline. Aerating the compost by turning should ensure an adequate supply of oxygen to the microbes.

Composting Period

The composting period is governed by a number of factors including, temperature, moisture, oxygen, particle size, the carbon-to-nitrogen ratio and the degree of turning involved. Generally, effective management of these factors will accelerate the composting process.

Carbon to Nitrogen Ratio

The microbes in compost use carbon for energy and nitrogen for protein synthesis. The proportion of these two elements required by the microbes averages about 30 parts carbon to 1 part nitrogen. Accordingly, the ideal ratio of Carbon to Nitrogen (C:N) is 30 to 1 (measured on a dry weight basis). This ratio governs the speed at which the microbes decompose organic waste.

Most organic materials do not have this ratio and, to accelerate the composting process, it may be necessary to balance the numbers.

The C:N ratio of materials can be calculated by using table below.

Example, if you have two bags of cow manure (C:N = 20:1) and one bag of corn stalks (C:N = 60:1) then combined you have a C:N ration of (20:1 + 20:1 + 60:1)/3 = (100:1)/3 = 33:1.

Table: lists the Carbon/Nitrogen Ratios of Some Common Organic Materials.

Material	C:N Ratio
Vegetable wastes	12-20:1
Alfalfa hay	13:1
Cow manure	20:1
Apple pomace	21:1
Leaves	40-80:1
Corn stalks	60:1
Oat straw	74:1
Wheat straw	80:1
Paper	150-200:1
Sawdust	100-500:1
Grass clippings	12-25:1
Coffee grounds	20:1
Bark	100-130:1
Fruit wastes	35:1
Poultry manure (fresh)	10:1
Horse manure	25:1
Newspaper	50-200:1
Pine needles	60-110:1
Rotted manure	20:1

The C:N ratios listed above are for guidelines only.

Composters for Smaller Volumes

Plastic bin (well ventilated) Metal or plastic drum (base removed – well ventilated).

Composters for Larger Volumes

Rotating drum (in vessel) Enclosure (made from 4 × 4 pallets lined with chicken wire) Open pile – windrow (covered with plastic or tarp).

In-Vessel

An in-vessel, aerobic mechanical composter can be constructed from a steel drum, or tank designed to rotate at three to five revolutions per hour. Rotation can be carried out with a simple hand crank or a timed electrical mechanical device. This type of composter can produce a stabilized compost in three to four days and can be an environmentally appropriate, low management alternative to bin composting.

Aerated Bin

An aerated bin can be constructed using 4 × 4 pallets fastened together to form a box and lined with wire mesh. To limit the degree of turning and permit air to flow through the pile the structure can be elevated or, in the alternative, perforated pipes can be incorporated into the structure. One side of the structure should be detachable to facilitate loading, mixing and unloading. The composter should be waterproof and located in and area that is protected from the wind.

Static compost piles and windrows should be large enough to retain heat and small enough to facilitate air to its center. As a rule of thumb, the minimum dimensions of a pile should be 3 feet by 3 feet by 3 feet.

Turning Units

Turning units are ideally suited for batch composting and are extremely practical for building and turning active compost. Turning units allow convenient mixing for aeration and accelerated composting.

Composting Methods

Hot Composting

Hot composting is the most efficient method for producing quality compost in a relatively short time. In addition, it favors the destruction of weed seeds, fly larvae and pathogens. While hot composting, using the windrow or bin method, requires a high degree of management, hot composting, using the in-vessel method, requires a lesser degree of management.

Cold Composting

This method is ideal for adding organic matter around trees, in garden plots, in eroded areas etc. The time required to decompose organic matter using this method is governed, to a large extent, by environmental conditions and could take two years or more.

Sheet Composting

Sheet composting is carried out by spreading organic material on the surface of the soil or untilled ground and allowing it to decompose naturally. Over time, the material will decompose and filter into the soil. This method is ideally suited for forage land, no-till applications, erosion control, roadside landscaping etc. The process does not favor the destruction of weed seeds, fly larvae, pathogens etc. and composting materials should be limited to plant residue and manure. Again, decomposition time is governed by environmental conditions and can be quite lengthy.

Trench Composting

Trench composting is relatively simple. Simply dig a trench 6 - 8 inches deep, fill with 3 - 4 inches of organic material and cover with soil. Wait a few weeks and plant directly above the trench. This method does not favor the destruction of weed seeds, fly larvae and pathogens and the composting process can be relatively slow.

Loading the Bin/Windrow

Place the raw materials in layers using a balance of high carbon (moist) and low carbon (dry) materials. Each layer should be no more than four to six inches in depth. This will initiate and accelerate the composting process and eliminate odors).

Procedure

Step 1: Start with a 4 to 6 inch layer of coarse material set on the bottom of the composter or on top of the soil.

Step 2: Add a 3 to 4 inch layer of low carbon material.

Step 3: Add a 4 to 6 inch layer of high carbon material.

Step 4: Add a 1 inch layer of garden soil or finished compost.

Step 5: Mix the layers of high carbon material, low carbon material, and soil or compost.

Repeat steps 2 through 5 until the composting bin is filled (maximum 4 feet in height). Cap with dry material.

Loading the Vessel (In-vessel Composting)

To accelerate the composting process, simply mix the high carbon and low carbon materials together before placing them in the composter. Add the mixture to the composter in small batches, spraying each batch with a light mist of water or CBCT stock solution.

Adding Material During the Composting Process

Ideally, new materials should be added to the composting system during turning or mixing. Generally, the addition of moist materials accelerates the composting process while the addition of dry materials slows the process.

Finished compost can be classified as a 100% organic fertilizer containing primary nutrients as well as trace minerals, humus and humic acids, in a slow release form. Compost improves soil porosity, drainage and aeration and moisture holding capacity and reduces compaction. Compost can retain up to ten times it's weight in water. In addition, compost helps buffer soils against extreme chemical imbalances; aids in unlocking soil minerals; releases nutrients over a wide time window; acts as a buffer against the absorption of chemicals and heavy metals; promotes the development of healthy root zones; suppresses diseases associated with certain fungi; and helps plants tolerate drought conditions.

Applications

Compost can be used in a variety of applications. High quality compost can be used in agriculture, horticulture, landscaping and home gardening. Medium quality compost can be used in applications such as erosion control and roadside landscaping. Low quality compost can be used as a landfill cover or in land reclamation projects.

Vermicomposting

Earthworms have been on the Earth for over 20 million years. In this time they have faithfully done their part to keep the cycle of life continuously moving. Their purpose is simple but very important.

They are nature's way of recycling organic nutrients from dead tissues back to living organisms. Many have recognized the value of these worms. Ancient civilizations, including Greece and Egypt valued the role earthworms played in soil. The Egyptian Pharaoh, Cleopatra said, "Earthworms are sacred." She recognized the important role the worms played in fertilizing the Nile Valley croplands after annual floods. Charles Darwin was intrigued by the worms and studied them for 39 years. Referring to an earthworm, Darwin said, "It may be doubted whether there are many other animals in the world which have played so important a part in the history of the world." The earthworm is a natural resource of fertility and life.

Earthworms live in the soil and feed on decaying organic material. After digestion, the undigested material moves through the alimentary canal of the earthworm, a thin layer of oil is deposited on the castings. This layer erodes over a period of 2 months. So although the plant nutrients are immediately available, they are slowly released to last longer. The process in the alimentary canal of the earthworm transforms organic waste to natural fertilizer. The chemical changes that organic wastes undergo include deodorizing and neutralizing. This means that the pH of the castings is 7 (neutral) and the castings are odorless. The worm castings also contain bacteria, so the process is continued in the soil, and microbiological activity is promoted.

Sieved finished vermicompost   Vermicompost ready for sale

Vermicomposting is the process of turning organic debris into worm castings. The worm castings are very important to the fertility of the soil. The castings contain high amounts of nitrogen, potassium, phosphorus, calcium, and magnesium. Castings contain: 5 times the available nitrogen, 7 times the available potash, and 1½ times more calcium than found in good topsoil. Several researchers have demonstrated that earthworm castings have excellent aeration, porosity, structure, drainage, and moisture-holding capacity. The content of the earthworm castings, along with the natural tillage by the worms burrowing action, enhances the permeability of water in the soil. Worm castings can hold close to nine times their weight in water. "Vermiconversion," or using earthworms to convert waste into soil additives, has been done on a relatively small scale for some time. A recommended rate of vermicompost application is 15-20 percent.

Vermicomposting is done on small and large scales. In the 1996 Summer Olympics in Sydney, Australia, the Australians used worms to take care of their tons and tons of waste. They then found that waste produced by the worms was could be very beneficial to their plants and soil. People in the U.S. have commercial vermicomposting facilities, where they raise worms and sell the castings that the worms produce. Then there are just people who own farms or even small gardens, and they may put earthworms into their compost heap, and then use that for fertilizer.

Vermicompost and its Utilization

Vermicompost is nothing but the excreta of earthworms, which is rich in humus and nutrients. We can rear earthworms artificially in a brick tank or near the stem/trunk of trees (specially horticultural trees). By feeding these earthworms with biomass and watching properly the food (bio-mass) of earthworms, we can produce the required quantities of vermicompost.

Materials for Preparation of Vermicompost

Any types of biodegradable wastes:

1. Crop residues.

2. Weed biomass.

3. Vegetable waste.

4. Leaf litter.

5. Hotel refuse.

6. Waste from agro-industries.

7. Biodegradable portion of urban and rural wastes.

Phase of Vermicomposting

Phase 1: Processing involving collection of wastes, shredding, mechanical separation of the metal, glass and ceramics and storage of organic wastes.

Phase 2: Pre digestion of organic waste for twenty days by heaping the material along with cattle dung slurry. This process partially digests the material and fit for earthworm consumption. Cattle dung and biogas slurry may be used after drying. Wet dung should not be used for vermicompost production.

Phase 3: Preparation of earthworm bed. A concrete base is required to put the waste for vermicompost preparation. Loose soil will allow the worms to go into soil and also while watering, all the dissolvable nutrients go into the soil along with water.

Phase 4: Collection of earthworm after vermicompost collection. Sieving the composted material to separate fully composted material. The partially composted material will be again put into vermicompost bed.

Phase 5: Storing the vermicompost in proper place to maintain moisture and allow the beneficial microorganisms to grow.

The Five Essentials need of Worms

Compost worms need five basic things:

1. An hospitable living environment, usually called "bedding".

2. A food source.

3. Adequate moisture (greater than 50% water content by weight).

4. Adequate aeration.

5. Protection from temperature extremes.

These five essentials are discussed in more detail below:

Bedding

Bedding is any material that provides the worms with a relatively stable habitat. This habitat must have the following characteristics:

High Absorbency

Worms breathe through their skins and therefore must have a moist environment in which to live. If a worm's skin dries out, it dies. The bedding must be able to absorb and retain water fairly well if the worms are to thrive.

Good Bulking Potential

If the material is too dense to begin with, or packs too tightly, then the flow of air is reduced or eliminated. Worms require oxygen to live, just as we do. Different materials affect the overall porosity of the bedding through a variety of factors, including the range of particle size and shape, the texture, and the strength and rigidity of its structure. The overall effect is referred to in this document as the material's bulking potential.

Low Protein and Nitrogen Content (High Carbon: Nitrogen Ratio)

Although the worms do consume their bedding as it breaks down, it is very important that this be a slow process. High protein/nitrogen levels can result in rapid degradation and its associated heating, creating inhospitable, often fatal, conditions. Heating can occur safely in the food layers of the vermiculture or vermicomposting system, but not in the bedding.

Requirements

1. Housing: Sheltered culturing of worms is recommended to protect the worms from excessive sunlight and rain. All the entrepreneurs have set up their units in vacant cowsheds, poultry sheds, basements and back yards.

2. Containers: Cement tanks were constructed. These were separated in half by a dividing wall. Another set of tanks were also constructed for preliminary decomposition.

3. Bedding and feeding materials: During the beginning of the enterprises, most women used cowdung in order to breed sufficient numbers of earthworms. Once they have large populations, they can start using all kinds of organic waste. Half of the entrepreneurs have now reached populations of 12,000 to 15,000 adult earthworms.

Vermicompost Production Methodology

Selection of Suitable Earthworm

For vermicompost production, the surface dwelling earthworm alone should be used. The earthworm, which lives below the soil, is not suitable for vermicompost production. The African earthworm (Eudrillus engenial), Red worms (Eisenia foetida) and composting worm (Peronyx excavatus) are promising worms used for vermicompost production. All the three worms can be mixed together for vermicompost production. The African worm (Eudrillus eugenial) is preferred over other two types, because it produces higher production of vermicompost in short period of time and more young ones in the composting period.

  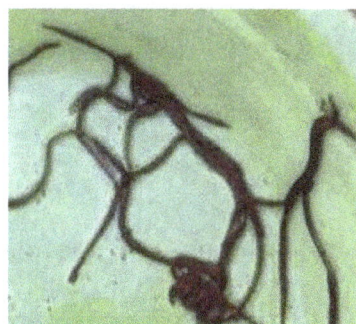

| African earthworm (Eudrillus euginiae) | Tiger worm or Red wrinkle (Eisenia foetida) | Asian worms (perinonyx ecavatus) |

Selection of Site for Vermicompost Production

Vermicompost can be produced in any place with shade, high humidity and cool. Abandoned cattle shed or poultry shed or unused buildings can be used. If it is to be produced in open area, shady place is selected. A thatched roof may be provided to protect the process from direct sunlight and rain. The waste heaped for vermicompost production should be covered with moist gunny bags.

Containers for Vermicompost Production

A cement tub may be constructed to a height of 2½ feet and a breadth of 3 feet. The length may be fixed to any level depending upon the size of the room. The bottom of the tub is made to slope like structure to drain the excess water from vermicompost unit. A small sump is necessary to collect the drain water.

In another option over the hand floor, hollow blocks/bricks may be arranged in compartment to a height of one feet, breadth of 3 feet and length to a desired level to have quick harvest. In this method, moisture assessment will be very easy. No excess water will be drained. Vermicompost can also be prepared in wooden boxes, plastic buckets or in any containers with a drain hole at the bottom.

Coir waste

Saw dust

Sugarcane trash

Cement tub

Vermiculture Bed

Vermiculture bed or worm bed (3 cm) can be prepared by placing after saw dust or husk or coir waste or sugarcane trash in the bottom of tub/container. A layer of fine sand (3 cm) should be spread over the culture bed followed by a layer of garden soil (3 cm). All layers must be moistened with water.

Common Bedding Materials

Bedding Material	Absorbency	Bulking Pot.	C:N Ratio
Horse Manure	Medium-Good	Good	22 - 56
Peat Moss	Good	Medium	58
Corn Silage	Medium-Good	Medium	38 - 43
Hay – general	Poor	Medium	15 - 32
Straw – general	Poor	Medium-Good	48 - 150
Straw – oat	Poor	Medium	48 - 98
Straw – wheat	Poor	Medium-Good	100 - 150
Paper from municipal waste stream	Medium-Good	Medium	127 - 178
Newspaper	Good	Medium	170
Bark – hardwoods	Poor	Good	116 - 436
Bark -- softwoods	Poor	Good	131 - 1285
Corrugated cardboard	Good	Medium	563
Lumber mill waste -- chipped	Poor	Good	170
Paper fibre sludge	Medium-Good	Medium	250
Paper mill sludge	Good	Medium	54
Sawdust	Poor-Medium	Poor-Medium	142 - 750
Shrub trimmings	Poor	Good	53
Hardwood chips, shavings	Poor	Good	451 - 819
Softwood chips, shavings	Poor	Good	212 - 1313
Leaves (dry, loose)	Poor-Medium	Poor-Medium	40 - 80
Corn stalks	Poor	Good	60 - 73

Corn cobs	Poor-Medium	Good	56 - 123
Paper mill sludge	Good	Medium	54
Sawdust	Poor-Medium	Poor-Medium	142 - 750
Shrub trimmings	Poor	Good	53
Hardwood chips, shavings	Poor	Good	451 - 819
Softwood chips, shavings	Poor	Good	212 - 1313
Leaves (dry, loose)	Poor-Medium	Poor-Medium	40 - 80
Corn stalks	Poor	Good	60 - 73
Corn cobs	Poor-Medium	Good	56 - 123

If available, shredded paper or cardboard makes an excellent bedding, particularly when combined with typical on-farm organic resources such as straw and hay. Organic producers, however, must be careful to ensure that such materials are not restricted under their organic certification standards. Paper or cardboard fibre collected in municipal waste programs cannot be approved for certification purposes. There may be cases, however, where fibre resources from specific generators could be sourced and approved. This must be considered on a case-by-case basis. Another material in this category is paper-mill sludge, which has the high absorbency and small particle size that so well complements the high C:N ratios and good bulking properties of straw, bark, shipped brush or wood shavings. Again, the sludge must be approved if the user has organic certification.

In general, it should be noted by the reader that the selection of bedding materials is a key to successful vermiculture or vermicomposting. Worms can be enormously productive (and reproductive) if conditions are good; however, their efficiency drops off rapidly when their basic needs are not met. Good bedding mixtures are an essential element in meeting those needs. They provide protection from extremes in temperature, the necessary levels and consistency of moisture, and an adequate supply of oxygen. Fortunately, given their critical importance to the process, good bedding mixtures are generally not hard to come by on farms. The most difficult criterion to meet adequately is usually absorption, as most straws and even hay are not good at holding moisture. This can be easily addressed by mixing some aged or composted cattle or sheep manure with the straw. The result is somewhat similar in its bedding characteristics to aged horse manure.

Mixing beddings need not be an onerous process; it can be done by hand with a pitchfork (small operations), with a tractor bucket (larger operations), or, if one is available, with an agricultural feed mixer. Please note that the latter would only be appropriate for large commercial vermicomposting operations where high efficiency levels and consistent product quality is required.

Worm Food

Compost worms are big eaters. Under ideal conditions, they are able to consume in excess of their body weight each day, although the general rule-of-thumb is ½ of their body weight per day. They will eat almost anything organic (that is, of plant or animal origin), but they definitely prefer some foods to others. Manures are the most commonly used worm feedstock, with dairy and beef manures generally considered the best natural food for Eisenia, with the possible exception of rabbit manure. The former, being more often available in large quantities, is the feed most often used.

Common Worm Feed Stocks

Food	Advantages	Disadvantages
Cattle manure	Good nutrition; natural food, therefore little adaptation required.	Weed seeds make pre-composting necessary
Poultry manure	High N content results in good nutrition and a high-value product.	High protein levels can be dangerous to worms, so must be used in small quantities; major adaptation required for worms not used to this feedstock. May be pre-composted but not necessary if used cautiously.
Sheep/Goat manure	Good nutrition	Require pre-composting (weed seeds); small particle size can lead to packing, necessitating extra bulking material.
Hog manure	Good nutrition; produces excellent vermi-compost.	Usually in liquid form, therefore must be dewatered or used with large quantities of highly absorbent bedding.
Rabbit manure	N content second only to poultry manure, there-fore good nutrition; contains very good mix of vitamins & minerals; ideal earth-worm feed.	Must be leached prior to use because of high urine content; can overheat if quantities too large; availability usually not good.
Fresh food scraps (e.g., peels, other food prep waste, leftovers, commercial food processing wastes).	Excellent nutrition, good moisture content, possibility of revenues from waste tipping fees.	Extremely variable (depending on source); high N can result in overheating; meat & high-fat wastes can create anaerobic conditions and odours, attract pests, so should NOT be included without pre-composting.
Pre-composted food wastes	Good nutrition; partial decomposition makes digestion by worms easier and faster; can include meat and other greasy wastes; less tendency to overheat.	Nutrition less than with fresh food wastes.
Biosolids (human waste)	Excellent nutrition and excellent product; can be activated or non-activated sludge, septic sludge; possibility of waste management revenues.	Heavy metal and/or chemical contamination (if from municipal sources); odour during application to beds (worms control fairly quickly); possibility of pathogen survival if process not complete.
Seaweed	Good nutrition; results in excellent product, high in micronutrients and beneficial microbes	Salt must be rinsed off, as it is detrimental to worms; availability varies by region.
Legume hays	Higher N content makes these good feed as well as reasonable bedding.	Moisture levels not as high as other feeds, requires more input and monitoring.
Legume hays	Higher N content makes these good feed as well as reasonable bedding.	Moisture levels not as high as other feeds, requires more input and monitoring.
Corrugated cardboard (including waxed).	Excellent nutrition (due to high-protein glue used to hold layers together); worms like this material; possible revenue source from WM fees.	Must be shredded (waxed variety) and/or soaked (non-waxed) prior to feeding.
Fish, poultry offal; blood wastes; animal mortalities.	High N content provides good nutrition; opportunity to turn problematic wastes into high-quality product.	Must be pre-composted until past thermophillic stage.

Selection for Vermicompost Production

Cattle dung (except pig, poultry and goat), farm wastes, crop residues, vegetable market waste, flower market waste, agro industrial waste, fruit market waste and all other bio degradable waste are suitable for vermicompost production. The cattle dung should be dried in open sunlight before used for vermicompost production. All other waste should be predigested with cow dung for twenty days before put into vermibed for composting.

Putting the Waste in the Container

The predigested waste material should be mud with 30% cattle dung either by weight or volume. The mixed waste is placed into the tub/container upto brim. The moisture level should be maintained at 60%. Over this material, the selected earthworm is placed uniformly. For one-meter length, one-meter breadth and 0.5-meter height, 1 kg of worm (1000 Nos.) is required. There is no necessity that earthworm should be put inside the waste. Earthworm will move inside on its own.

Watering the Vermibed

Daily watering is not required for vermibed. But 60% moisture should be maintained throughout the period. If necessity arises, water should be sprinkled over the bed rather than pouring the water. Watering should be stopped before the harvest of vermicompost.

Harvesting Vermicompost

In the tub method of composting, the castings formed on the top layer are collected periodically. The collection may be carried out once in a week. With hand the casting will be scooped out and put in a shady place as heap like structure. The harvesting of casting should be limited up to earthworm presence on top layer. This periodical harvesting is necessary for free flow and retain the compost quality. Otherwise the finished compost get compacted when watering is done. In small bed type of vermicomposting method, periodical harvesting is not required. Since the height of the waste material heaped is around 1 foot, the produced vermicompost will be harvested after the process is over.

Harvesting Earthworm

After the vermicompost production, the earthworm present in the tub/small bed may be harvested by trapping method. In the vermibed, before harvesting the compost, small, fresh cow dung ball is made and inserted inside the bed in five or six places. After 24 hours, the cow dung ball is removed. All the worms will be adhered into the ball. Putting the cow dung ball in a bucket of water will separate this adhered worm. The collected worms will be used for next batch of composting.

Worm harvesting is usually carried out in order to sell the worms, rather than to start new worm beds. Expanding the operation (new beds) can be accomplished by splitting the beds that is, removing a portion of the bed to start a new one and replacing the material with new bedding and feed. When worms are sold, however, they are usually separated, weighed, and then transported in a relatively sterile medium, such as peat moss. To accomplish this, the worms must first be separated from the bedding and vermicompost. There are three basic categories of methods used by growers to harvest worms: manual, migration, and mechanical.

Manual Methods

Manual methods are the ones used by hobbyists and smaller-scale growers, particularly those who sell worms to the home-vermicomposting or bait market. In essence, manual harvesting involves hand-sorting, or picking the worms directly from the compost by hand. This process can be facilitated by taking advantage of the fact that worms avoid light. If material containing worms is dumped in a pile on a flat surface with a light above, the worms will quickly dive below the surface. The harvester can then remove a layer of compost, stopping when worms become visible again. This process is repeated several times until there is nothing left on the table except a huddled mass of worms under a thin covering of compost. These worms can then be quickly scooped into a container, weighed, and prepared for delivery.

There are several minor variations and enhancements on this method, such as using a container instead of a flat surface, or making several piles at once, so that the person harvesting can move from one to another, returning to the first one in time to remove the next layer of compost. They are all labour-intensive, however, and only make sense if the operation is small and the value of the worms is high.

Self-Harvesting (Migration) Methods

These methods, like some of the methods used in vermicomposting, are based on the worms tendency to migrate to new regions, either to find new food or to avoid undesirable conditions, such as dryness or light. Unlike the manual methods described above, however, they often make use of simple mechanisms, such as screens or onion bags.

The screen method is very common and easy to use. A box is constructed with a screen bottom. The mesh is usually ¼", although 1/8" can be used as well. There are two different approaches. The downward-migration system is similar to the manual system, in that the worms are forced downward by strong light. The difference with the screen system is that the worms go down through the screen into a prepared, pre-weighed container of moist peat moss. Once the worms have all gone through, the compost in the box is removed and a new batch of worm-rich compost is put in. The process is repeated until the box with the peat moss has reached the desired weight. Like the manual method, this system can be set up in a number of locations at once, so that the worm harvester can move from one box to the next, with no time wasted waiting for the worms to migrate.

The upward-migration system is similar, except that the box with the mesh bottom is placed directly on the worm bed. It has been filled with a few centimeters of damp peat moss and then sprinkled with a food attractive to worms, such as chicken mash, coffee grounds, or fresh cattle manure. The box is removed and weighed after visual inspection indicates that sufficient worms have moved up into the material. This system is used extensively in Cuba, with the difference that large onion bags are used instead of boxes. The advantage of this system is that the worm beds are not disturbed. The main disadvantage is that the harvested worms are in material that contains a fair amount of unprocessed food, making the material messier and opening up the possibility of heating inside the package if the worms are shipped. The latter problem can be avoided by removing any obvious food and allowing a bit of time for the worms to consume what is left before packaging.

Nutritive Value of Vermicompost

The nutrients content in vermicompost vary depending on the waste materials that is being used for compost preparation. If the waste materials are heterogeneous one, there will be wide range of nutrients available in the compost. If the waste materials are homogenous one, there will be only certain nutrients are available. The common available nutrients in vermicompost is as follows:

- Organic carbon: 9.5 – 17.98%.

- Nitrogen : 0.5 – 1.50%.

- Phosphorous: 0.1 – 0.30%.

- Potassium: 0.15 – 0.56%.

- Sodium: 0.06 – 0.30%.

- Calcium and Magnesium: 22.67 to 47.60 meq/100g.

- Copper: 2 – 9.50 mg kg-1.

- Iron : 2 – 9.30 mg kg-1.

- Zinc: 5.70 – 11.50 mg kg-1.

- Sulphur: 128 – 548 mg kg-1.

Storing and Packing of Vermicompost

The harvested vermicompost should be stored in dark, cool place. It should have minimum 40% moisture. Sunlight should not fall over the composted material. It will lead to loss of moisture and nutrient content. It is advocated that the harvested composted material is openly stored rather than packed in over sac. Packing can be done at the time of selling. If it is stored in open place, periodical sprinkling of water may be done to maintain moisture level and also to maintain beneficial microbial population. If the necessity comes to store the material, laminated over sac is used for packing. This will minimize the moisture evaporation loss. Vermicompost can be stored for one year without loss of its quality, if the moisture is maintained at 40% level.

Advantages of Vermicompost

- Vermicompost is rich in all essential plant nutrients.

- Provides excellent effect on overall plant growth, encourages the growth of new.

- Shoots leaves and improves the quality and shelf life of the produce.

- Vermicompost is free flowing, easy to apply, handle and store and does not have bad.

- Odour.

- It improves soil structure, texture, aeration, and waterholding capacity and prevents.

- Soil erosion.

- Vermicompost is rich in beneficial micro flora such as a fixers, P- solubilizers,

- Cellulose decomposing micro-flora etc in addition to improve soil environment.

- Vermicompost contains earthworm cocoons and increases the population and

- Activity of earthworm in the soil.

- It neutralizes the soil protection.

- It prevents nutrient losses and increases the use efficiency of chemical fertilizers.

- Vermicompost is free from pathogens, toxic elements, weed seeds etc.

- Vermicompost minimizes the incidence of pest and diseases.

- It enhances the decomposition of organic matter in soil.

- It contains valuable vitamins, enzymes and hormones like auxins, gibberellins etc.

Pests and Diseases of Vermicompost

Compost worms are not subject to diseases caused by micro-organisms, but they are subject to predation by certain animals and insects (red mites are the worst) and to a disease known as "sour crop" caused by environmental conditions.

Organic Phosphorus Fertilizers

Organic phosphorus fertilizers come primarily from mineral sources, like rock dust or colloidal phosphate (also called "soft phosphate"), or from bone sources, such as steamed bone meal or fish bone meal.

Mineral phosphorus sources are cheaper and last longer in the soil. Bone sources are more readily absorbed by plants.

Phosphorus is needed for root development, stem formation, and fruiting in summer vegetables like tomatoes, peppers, eggplants, squash, melons, and cucumbers.

Phosphorus tends to be widely disbursed in soil, so it's hard for these plants to get enough of it within their limited root zones. To get enough phosphorus to produce fruit, fruiting plants evolved symbiotic relationships with myccorhizal fungi. Almost all plants that bear fruit form myccorhizal associations.

Fungi are creatures of the soil. Their hyphae can spread for hundreds of feet underground (the largest living organisms are fungi), and they can transport nutrients anywhere in the hyphal system. Myccorhizal fungi concentrate phosphorus and other minerals at the roots of plants, and the plants provide the fungi with sugars, starches, and amino acids in exchange.

Gardeners and farmers usually add supplemental organic phosphorus fertilizers to the soil to accommodate crop needs. Inoculating seedling roots with Endo-Myccorhizae increases their ability to absorb soil phosphorus.

It's especially useful for growing vegetables in containers, where sterile potting mixes limit plant growth.

The table below lists organic phosphorus fertilizers. Colloidal phosphate is more biologically available than rock dust, but not as readily assimilated by plants as bone sources of organic phosphorus.

Organic Phosphorus Fertilizers (P)—Links Go to Offsite Affiliates to Purchase Organic Soil Amendments				
Soil Amendment	N-P-K	Description	Lasts	Application Rate
t Rock Phosphate	0-18-0	Colloidal Phosphate has a clay base that makes it easier for plants to assimilate than phosphate rock. Releases over months and years in acidic and neutral soils, but breaks down poorly in alkaline soils (pH higher than 7). Peak availability in 2nd year.	2-3 Years	Up to 6lbs/100 sq ft

Bat Guano (High-P)	0-5-0	High-Phosphate guano from fruit-eating bats. Excellent P source for container vegetables and gardens.	2-3 Years	2-3lbs/100 sq ft
Steamed Bone Meal	3-15-0	Made from ground cattle bones. P in bone meal is highly plant-available. Great mixed into the planting hole with bulbs. Good amendment for allium family plants (onions, garlic). May attract raccoons. P in bone meal not released in alkaline (pH greater than 7) soils.	1-4 Months	10lbs/100 sq ft
Fish Bone Meal	3-18-0	Phosphorus from fish bone meal is readily assimilated by microorganisms and plant roots in the soil.	1-2 Years	1-2lbs/100 sq ft
Rock Phosphate	0-33-0	Veryslow release P source. Releases over several years in acidic and neutral soils, but won't break down in alkaline soils (pH higher than 7).	3-5 Years	Up to 6lbs/100 sq ft.
Rock Dust (Crushed Granite)	0—3-5—0, trace minerals	Granite fines, the dust from rock grinding and sorting operations.Veryslow releasing P source, good source of trace minerals for plant immunity and tolerance of temperature extremes.	5-10 Years	Up to 8.5lbs/100 sq ft
Chicken Manure	1.1-0.8-0.5	Good manure source for P and some K.	3-12 Months	1/2-1" layer (5-10 5-gal buckets/100 sq ft)
Pig Manure	0.8-0.7-0.5	Good, balanced manure source of N, P, and K. Because some pig parasites and pathogens can infect humans, pig manure is not allowed in many organic protocols. If it is used, itmustbe hot-composted prior to use.	3-12 Months	1" layer (10 5-gal buckets/100 sq ft)

Phosphorus Sources and Management in Organic Production Systems

Organically produced fruit and vegetables are among the fastest growing agricultural markets. With greater demand for organically grown produce, more farmers are considering organic production options. Furthermore, there is an increasing interest in maintaining optimal production in an organic system, which involves appropriate nutrient management. Organic production systems seek to improve soil organic matter and biological diversity, which may impact P cycling and P uptake by crops. Increases in organic matter will be accompanied by an increase in the organic P pool. Furthermore, management of cover crops and potentially enhanced arbuscular mycorrhizal fungi colonization from organic production practices can increase the availability of soil P pool (both organic and inorganic) by stimulating microbial activity and release of root exudates. This can help compensate for low soil P, but will not supersede the need to replace P removed by the harvested crop. Phosphorus fertilization in organic production systems entails balancing the P inputs with crop removal through selection and management of both nitrogen (N) and P inputs. Organic production systems that rely on manure or composts for meeting crop N demand will likely have a P surplus; therefore, P deficiencies will not be an issue. Systems using other N sources may have a P deficit, therefore requiring P supplementation for optimal plant growth. In such situations, maintenance P applications equal to crop removal should be made based on soil test recommendations. Primary organically approved P sources are phosphate rock (PR), manure, and compost. Phosphate rock is most effective at supplying P in soils with low pH (less than 5.5) and low calcium concentrations. Phosphate rock applications made to soils with pH greater than 5.5 may not be effective because of reduced PR solubility. Manure- and compost-based P has high plant availability, ranging from 70% to 100% available. Use of manures and composts requires extra considerations to reduce the risk of P loss from P sources to surface waters. Best management practices (BMPs) for reducing source P losses are incorporation of the manures or composts and timing applications to correspond to periods of low runoff risk based on climatic conditions. Organic production systems that use manures and composts as their primary N source should focus on minimizing P buildup in the soils and use of management practices that reduce the risks of P loss to surface waters. Evaluation of P loss risk with a P index will assist in identification of soil and management factors likely to contribute to high P loss as well as BMPs that can decrease P loss risks. BMPs should focus on controlling both particulate and dissolved P losses.

Philosophies of nutrient management in organic production systems focus on maintaining agricultural productivity with minimal inputs. The end goal of nutrient management in organic agriculture is to produce food in a more environmentally sustainable system that takes advantage of internal nutrient cycling and reduces losses. Nutrient inputs to organic production systems are focused on carbon-based nutrient sources (e.g., crop residue, compost, manure) and nonprocessed mineral sources (e.g., rock phosphate, lime, gypsum). As such, nutrient management in organic production systems is fundamentally different from that in conventional systems. Phosphorus (P) management is of particular interest because the P sources approved for use in organic agriculture have diverse characteristics. Phosphorus management can also have a strong influence on the environmental impact of crop production because P is a leading contributor to water quality degradation. Furthermore, nitrogen (N) management decisions in organic production often influence P availability for crop use and potential risks of P loss to the environment.

Units to convert U.S. to SI, Multiply by	U.S. Unit	SI Unit	To convert SI to U.S., Multiply by
1.1209	Lb/acre	$Kg.ha^{-1}$	0.8922
1	micron	μm	1
1	Ppm	$Mg.kg^{-1}$	1
1	Ppm	$Mg.L^{-1}$	1
2.2417	ton/acre	$Mg.ha^{-1}$	0.4461

Phosphorus is an essential element for plant growth and is involved in many plant metabolic functions. Most notably, P is an essential component of adenosine diphosphate and adenosine triphosphate—organic molecules that are used for energy storage and transfer. Phosphorus is also a structural component of nucleotides, phospholipids, phosphoproteins, and coenzymes. Therefore, sustainable agricultural production is dependent on the maintenance of adequate P availability by the soil or other P inputs to the system.

Compared with undisturbed ecosystems, agricultural systems short-circuit complete P cycling because P in harvested crops is removed from fields, introduced into the human food chain, and processed through the waste stream. This open cycle creates a P deficit in soils without P additions. Several studies investigating whole-farm P budgets have found annual P deficits in organic production systems and conclude that these systems are mining P reserves built up from previous P inputs when soils were under conventional management. As illustrated in figure, long-term research studies have found drastic declines in production capacity for systems lacking P inputs. Because P is an essential nutrient for plant growth, sustainable systems should at a minimum replace the P removed in harvested crops to avoid such yield declines. Although organic agriculture seeks to maintain minimal inputs, it is advised that producers replace P removed in harvested crops.

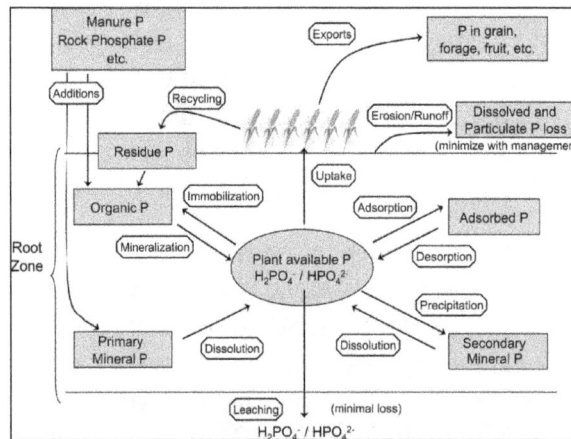

The phosphorus (P) cycle in organic production systems, including P additions, exports, cycling, and transformations within the soil

Plant-available P is present as inorganic orthophosphate species HPO_4^- and $H_2PO_4^-$. Phosphorus cycles in conventional systems would have different additions (chemical fertilizers) that would directly supply plant-available P. Phosphorus cycling in unmanaged ecosystems would have no harvest removal and no P additions.

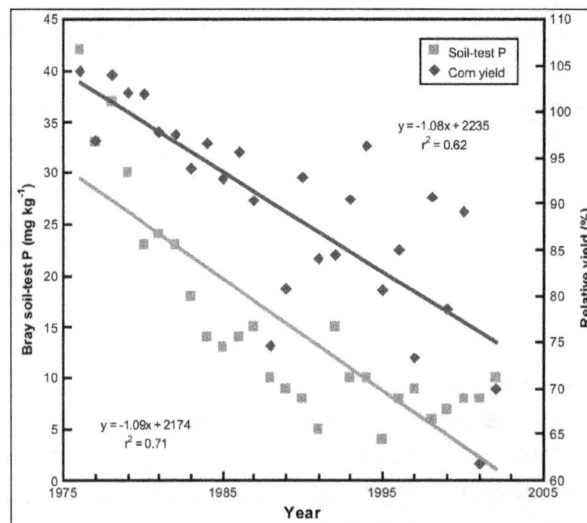

Decline in soil test phosphorus (P) and relative corn yield after 27 years of continuous cropping without addition of P fertilizer. Yield is relative to corn plots receiving 33 kg·ha−1 (29.4 lb/acre) P per year.

Although some organic farms may run a P deficit, others are likely to have a P surplus because many organic production systems rely on manures or composts as N sources. According to a poll conducted by the Organic Farming Research Foundation, compost and manure applications are regularly made by 57% and 22% of U.S. organic producers, respectively. Farms that use compost or manure to meet crop N requirements will generally have a P surplus. Although P requirements for crop growth are not a concern in these management systems, sustainable P management must include measures to minimize P losses to surface waters.

Phosphorus management in organic production systems should therefore consider cropping system effects on P availability, N amendments and concomitant P applications, P sources and availability, and best management practices to reduce P losses.

Organic Agriculture Effects on Soil Properties and Phosphorus Availability

Soil processes that affect P cycling and availability in organically managed soils are not different from those in conventionally managed soils, but the relative importance of P cycling may differ between the two systems. Organic production potentially affects a number of soil properties, including soil organic matter content, microbial activity, microbial community structure, soil aggregation, water-holding content, and soil chemistry, which could potentially affect P availability. Organic production systems often contain more diverse crop rotations, including cover crops, which may also alter nutrient cycling. The primary means by which organic production practices would influence P availability would be through alteration of soil organic matter content, increased P availability from degradation of cover crop residues, and increased arbuscular mycorrhizal fungi (AMF) colonization resulting from lack of soluble P fertilizer applications.

Soil Organic Matter

Many long-term experiments comparing conventional and organic practices have documented increased organic matter/carbon accumulation in organically managed soils. However, some long-term research experiments and on-farm comparisons did not find increases in soil organic matter resulting from organic production practices. Whether there is in fact any increase at all depends on the specific practices used in the organic and conventional systems (crop residue return, manure and compost application, tillage, and so on). Several studies have shown an increase in the concentration of total P in organic production systems; however, other studies have shown a decrease in the mineral or available P pools.

Increased soil organic matter content could increase plant P uptake by decreasing bulk density and increasing porosity, thereby improving root exploration and effectiveness. In addition, the organic P pool increases with increasing organic matter. Organic P compounds such as inositol phosphates, nucleic acids, and phospholipids present in organic matter can be mineralized during organic matter decomposition, thereby increasing P availability and acting as a P source for future crops. Increasing the availability of the organic matter P pool has been the subject of considerable research on both cover crops and AMF.

Cover Crops

Cover crops grown before cash crops take up soil P and then release it as the cover crop decomposes. In an incubation study, a legume amended soil released 0.27 mg·kg^{-1} P per day over 21 d versus 0.06 mg·kg^{-1} P per day for an unamended soil. In addition, some cover crops release root exudates such as carboxylates, enabling uptake of soil P that is not available to other plants. However, when various legume cover crops have been compared for their effectiveness in making P available to subsequent crops, the results have not always been predictable.

In a pot experiment using soils from Western Australia, fava bean (Vicia faba) promoted more growth and P uptake in subsequent wheat (Triticum aestivum) than the other legumes in the experiment, including white lupin (Lupinus albus) and field pea (Pisum sativum), which both had larger amounts of rhizosphere carboxylates than the fava bean. The P content of the fava bean was significantly higher than the other legumes at all soil P levels, or almost double, on a per-pot basis despite the lack of rhizosphere carboxylates. the P benefits to the wheat in their experiment were the result of the mineralization of the organic P from the fava bean rather than carboxylate-induced changes in soil chemistry.

Cavigelli and Thien found in a greenhouse pot study that sorghum (Sorghum bicolor) P uptake was positively correlated with P uptake of the preceding perennial forage cover crops [alfalfa (Medicago sativa), red clover (Trifolium pratense), and sweet clover (Melilotus officinalis)] but not correlated with P uptake of winter annual cover crops. For example, white lupin biomass and P uptake were two to three times greater than the other winter cover crops [austrian winter pea (Pisum sativum ssp. sativum var. arvense), vetch (Vicia villosa), and wheat], and yet sorghum P uptake after lupin was lower than all other treatments, including the control. In this case, cover crop species rather than P uptake seemed to influence subsequent sorghum P uptake. Furthermore, they found that cover crops that resulted in greater P uptake by the subsequent sorghum crop also caused greater declines in soil test P, thus illustrating that although cover crops may increase P uptake, it may be at the expense of other P pools that will need to be replenished to maintain production. They also suggest that a soil test measuring microbial activity and other changes in soil characteristics may be more useful for measuring P availability after incorporation of legumes than traditional soil testing (e.g., the Bray P1 test).

In a West African sorghum cropping system, crotalaria (Crotalaria retusa) used as a green manure was superior to cowpea (Vigna unguiculata), and both resulted in more yield from a subsequent corn (Zea mays) crop than sorghum in rotation. However, all treatments responded to additional single superphosphate [$Ca(H_2PO_4)_2 \cdot H_2O$ (0N–7.9P–0K)] applications, demonstrating that the green manure cover crops did not supply enough P to overcome P limitation. Similar results were obtained by Horst et al. (2001) in testing 16 legume cover crops in a corn-based cropping system in Nigeria. On one of the two soils tested, corn P uptake was correlated with phosphate application in the preceding cover crops. The cover-cropped soils also had higher rates of AMF infection.

These results indicate that total uptake of P by a prior cover crop is important, but at the same time, some responses, both positive and negative, seem to be species-specific and may be microbially mediated. A closer look at the relationship among green manures, P, and root colonizing mycorrhizal fungi may help explain some of these relationships.

Arbuscular Mycorrhizal Fungi Colonization

Enhanced P uptake is generally understood to be one of the most important benefits of AMF colonization. In return, the host plant provides carbohydrates usually, but not always, without detrimental effects to the host. Additional benefits of AMF colonization include increased tolerance to water stress and occasional suppression of crop pests and diseases, although this is less consistently observed than effects on nutrient uptake. Arbuscular mycorrhiza fungi are also important in helping to form water-stable aggregates, an indicator of soil quality. Over 80% of plant species can serve as AMF hosts. Notable exceptions sometimes used as crops or rotation crops in organic systems include plants in the Brassiceae and the Chenopodiaceae families.

Many practices used by organic farmers promote AMF, including the use of cover crops, and reduced use of biocides and soluble fertilizers, although the effect of any particular biocide may be difficult to predict and is not always negative. Reduced tillage also leads to higher rates of colonization with AMF, probably as a result of lack of disruption of the common mycorrhizal hyphae network. Frequent or deep tillage used by many organic farmers for weed control can reduce AMF abundance as can bare or fallow soil. The degree and frequency of tillage will depend on the cropping system. Rotation with perennial legumes could potentially compensate for some of the AMF reduction occurring during parts of the rotation cycle with frequent tillage.

AMF colonization appears to be greater in low P soils rather than high, although application of P in the form of manures and composts does not always inhibit colonization. An organic system with animal manure additions in a long-term Pennsylvania experiment had higher levels of soil P and also higher populations of AMF spores and colonized roots as compared with a conventional cropping system.

In a greenhouse study with leeks (Allium porrum), AMF increased P uptake from bone meal by 62% but did not enhance P uptake from Kola apatite, a form of igneous rock phosphate. Inoculation with AMF in a field trial increased growth and P uptake by leeks on a conventional soil, but had a slight negative effect on leeks on an organically managed soil, which had higher P and AMF levels compared with the conventional soil at the initiation of the experiment. Their conclusion was that working with indigenous AMF through systems management on organic farms has more potential than inoculation.

In summary, although AMF can help compensate for low P in some organic systems, even with good colonization, this does not always translate into higher yields. However, other benefits can accrue from AMF colonization, even with relatively high soil P levels, and organic farmers will often be using practices that promote higher AMF levels compared with conventional farmers. If soils are low in P, both cover crops and AMF can help increase uptake of P that is present, but farmers are advised to supplement with forms of P allowable under organic certification standards to meet overall crop needs and maintain a sufficient P balance in the system.

Nitrogen Source Effects on Phosphorus

Nitrogen sources used for organic production can have a strong impact on P availability, field-level P balance, and P management strategies. Animal manures and composts contain both N and P; however, the available N:P ratios in most manures and composts are less than that required by plants. For example, the average N:P ratio in the harvested portion of crop biomass is 7:1 with a minimum of 4:1; however, the N:P ratio of animal manures is frequently less than 4:1. Because not all N in manure is available for crop uptake, manure-based N application rates are calculated with potentially available N (PAN) concentrations in manure. Nitrogen availability of manure can range from 25% to 70% of total N; therefore, PAN:P ratios would be 30% to 75% less than the N:P ratios presented in table depending on manure characteristics and methods of application. This illustrates that the use of manure to meet crop N requirement will generally oversupply P, exceeding crop removal by as much as four times. Overapplication of P is not detrimental to crop growth or yields, but excess P applications will increase soil test P levels far beyond crop requirements. Both high P application rates and high soil test P can increase the risk of environmental damage resulting from P losses to surface waters.

Table: Average yields, nitrogen (N) removal, phosphorus (P) removal, and N:P ratios for harvested portions of common grain, forage, fruit, and vegetable crops.

Crop (common name)	Genus and species	Yield (Mg-ha-1)Y	N removal (kg-ha-1)`'	P removal (kg-ha-1)	N:P ratio
Grain crops					
Barley	Hordeum vulgarL	2.2	39.2	7.3	5.3
Corn	Zea mays	9.4	151.3	25.9	5.8
Oat	Avena sativa	2.9	56.0	9.8	5.7
Rice	Oryza sariva	4.0	56.0	9.8	5.7
Rye	Secale cereale	1.9	39.2	4.9	8.0
Sorghum	Sorghum rulgare	3.8	56.0	12.2	4.6
Wheat	Trificum acstivum	2.7	56.0	12.2	4.6
Bean, dry	Phaseolus vulgaris	2.0	84.0	12.2	6.9
Soybean	Glycine max	2.7	168.1	17.1	9.8
Forage crops					
Alfalfa	Medicago sativa	9.0	201.7	19.6	10.3
Bluegrass	Pao anntitr	4.5	67.2	9.8	6.9
Coastal bermuda	Cynodon dactylon	17.9	336.2	34.3	9.8
Cowpea	Vtana unguicalata	4.5	134,5	12.2	11.0
Peanut	Arachis hypogaea	5.0	117.7	12.2	9.6
Red clover	Trifolium pretense	5.6	112.1	12.2	9.2
Soybean	Glyeine max	4.5	100.9	9.8	10.3
Timothy	Ma'am pretense	5.6	67.2	12.2	5.5
Fruit and vegetable crops					
Apple	Mains pumila	26.9	33.6	4.9	6.9
Orange	Citrus sinensis	62.8	95.3	14.7	6.5
Peach	Prunus persica	33.6	39.2	9.8	4.0
Cabbage	Brassica oleracea var. capitata	44.8	145.7	17.1	8.5
Onion	Alliam cepa	16.8	50.4	9.8	5.2
Potato	Sulanaarn tilberiblfrn	26.9	89.7	14.7	6.1
Spinach	Spinacia oieracca	11.2	56.0	7.3	7.6
Sweet potato	rpomoca batatas	18.5	50.4	7.3	6.9
Tomato	Lyropersicon esculentum	44.8	134.5	19.6	6.9
Turnip	Brassica raga var. rapiftra	22.4	50.4	9.8	5.2

'Eakin, 1976.

I Mg•ha $^{-1}$ = 0.4461 tom/acre, 1 kg.ha^{-1} = 0.8922 Ib/acre.

Table: Average nitrogen (N) and phosphorus (P) concentrations reported for various manures and composts.

Nutrient source	n	N P (g.kg⁻¹)		N:P ratio Maximum avg minimum			References[y]
Manures							
Beer cattle	11	15.5	4.5	5.4	3.6	2.0	1,2,3,4
Dairy	16	34.4	8.4	6.6	4.2	1.7	5,6,7,8
Poultry	14	38.2	21.0	2.9	1.9	1.3	6,7,9,10,11,12
Swine	7	39.7	14.0	4.3	2.7	2.0	2,6,7,13
Cornposts							
Ileercattle	15	15.5	7.0	3.6	2.3	1.7	1,3,4,14
Dairy	8	25.5	13.5	3.2	2.0	0.7	14
Poultry	6	23.8	20.3	1.8	1.1	0.6	9,10,14,15
Swine	7	14.0	7.6	2.4	1.9	1.4	13,16
Yard waste	25	14.6	2.5	12,9	6.7	1.1	10,14,17,18,19,20
Yard waste+	5	15.4	7.9	2.4	1.6	0.4	21,22

Phosphorus and N concentrations of manures and composts vary dramatically based on the animal species, feeding regime, processing, and composting substrates. In general, poultry litters have higher P concentrations and lower N:P ratios than other manures. Manure from organically raised livestock may have lower P concentrations, but PAN:P ratios would still be less than what would be required by crops. Compost is more frequently applied in organic production than is manure because nutrients in compost are more stable and believed to be more environmentally benign uring the composting process, whereas some of the N is lost. Furthermore, composts have lower N availability than manures, exacerbating the disparity between the PAN:P ratio of compost versus N:P ratio in harvested crop biomass. However, yard waste compost (without added manure) has the highest N:P ratio of the organically based N sources listed in table. Although use of yard waste compost as a primary N source could help balance P inputs with P removal, its availability and relatively low N concentration would limit its widespread use.

Organic producers using manures and composts as a major N source will likely have high soil test P and their P management strategies should focus on reducing P losses to surface waters. On the other hand, organic producers who use manures and composts sparingly or not at all will likely have greater crop P removal than P inputs and may need supplemental P sources to sustain adequate crop production. Soil testing and knowledge of management history will be the best tools for indicating if P is a limiting nutrient in the production system.

Phosphorus Sources in Organic Agriculture

Phosphorus sources approved for use in organic agriculture have diverse properties that affect P availability and management. Common P sources include rock phosphate, manure, and compost, all of which are frequently used in research studies. Bone meal and guano are among the less commonly cited P sources but can have high P contents (ranging from 7% to 12% and 1% to 9%, respectively).

Phosphate Rock

Direct application of phosphate rock (PR) to soils as a P fertilizer has been practiced for over 100 years. Because direct application of PR continues to be an important P source in developing nations, there is a wealth of research addressing soil, crop, and PR effects on P availability, including three extensive reviews compiled during the last 30 years Phosphate rock is a slowly soluble P source. Although the total P concentration can be relatively high (greater than 15%), the soluble P concentration can be very low (less than 1%). Therefore, a few basic issues must be considered when evaluating the use of PR as a direct P source in agriculture, including PR properties, soil properties, climate, crop species, and soil management practices.

Although the phosphate compound found in PR is always some form of the mineral apatite [$Ca_5(PO_4)_3X$, where X is an anion], the chemical and mineralogical properties, and therefore solubility, vary greatly between PR sources. Soluble P concentration of PR is determined and expressed as water- and citrate-soluble P, similar to the methods used for conventional P fertilizers. Phosphate rocks with greater soluble P concentrations will generally have greater agronomic effectiveness. However, both the total P concentration and the soluble P concentration should be considered when deciding on PR sources. Particle size also affects PR reactivity, in which decreasing particle size down to ≈150 μm increases PR effectiveness. Decreasing PR particle size to less than 150 μm can be cost-prohibitive and does not result in additional agronomic benefit.

Phosphate rock sources can be generally classified as either sedimentary or igneous. Sedimentary PR has higher carbonate substitution and up to 20 times greater specific surface area than igneous rocks. Increases in carbonate substitution and specific surface area increase P solubility and make sedimentary PR sources better suited to direct application to soils. Phosphate rock sources known to have consistently high P availability are located in North Carolina and Gafsa, Tunisia; however, North Carolina PR is only used for production of processed phosphate concentrates and cannot be obtained as a raw phosphate ore now. Phosphate rock from the western United States has among the lowest solubility of PR worldwide. Although the PR source mine and mineralogy have a strong influence on PR solubility and P availability, this information can be difficult to obtain for PR sold on the retail market.

Table: Total and citrate-soluble phosphorus (P) concentrations in phosphate rock (PR) from various sources.

PR source	PR type	Total P	Citrate soluble P (g.kg⁻¹)[z]	Citrate soluble P (% total P)
Algeria[y]	Sedimentary	131	48	37
Gafsa,Tunisia[x]	Sedimentary	127	23	18
North Carolina[x]	Sedimentary	117	20	17
Florida[y]	Sedimentary	157	13	8
Tennessee[w]	Sedimentary	131	11	9
Montane[y]	Igneous	159	8	5
Araxa, Brazil[w]	Igneous	162	6	4

[x]1.g.kg⁻¹= 1000 ppm.

[y]Zal-iarah and Bah.

[x]Centre for Industrial Development.

[w]Van Kauwenbergh and McClellan.

Soil pH is one of the primary soil factors affecting PR efficacy. When PR is applied to soils, the apatite dissolves according to the reaction shown in Equation,

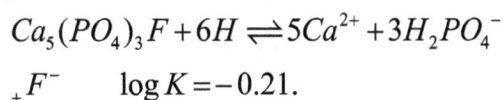

$$Ca_5(PO_4)_3 F + 6H \rightleftharpoons 5Ca^{2+} + 3H_2PO_4^-$$
$$_+ F^- \quad \log K = -0.21.$$

Although above equation is for fluorapatite, the general expression applies equally to other apatite minerals. As can be seen in Eq. 1, an increase in the H+ concentration (decrease in pH) shifts the reaction toward the reactants and increases phosphate concentration. By fixing the Ca2+ and F– activities at 10−2.5 and 10−4 mol·L−1, respectively, the phosphate concentration can be determined as a function of pH as follows:

$$\log (H_2PO_4^-) = 5.43 - 2pH.$$

Equation 2 illustrates that when other variables are constant, each unit increase in pH decreases phosphate concentration by two orders of magnitude. For example, phosphate concentration from fluorapatite dissolution would decrease from 0.8 mg·L−1 at pH 5 to 0.008 mg·L−1 at pH 6.

Because Ca^{2+} is a reaction product in apatite dissolution, high soil Ca^{2+} concentrations reduce apatite efficacy as a P source and soil properties that remove Ca^{2+} from soil solution will increase dissolution. Soils with high cation exchange capacity and low Ca^{2+} saturation tend to maintain low Ca^{2+} concentrations in soil solution, thereby promoting apatite dissolution. High leaching rates on soils with low cation exchange capacity can also increase PR dissolution because the excess water will move Ca^{2+} and other reaction products away from the vicinity of the dissolving apatite. Increasing organic matter content of the soil can also improve the effectiveness of PR as a P source because organic matter increases cation exchange capacity, organic acids form complexes with free Ca^{2+}, and organic matter increases titratable acidity.

It is difficult to make universally applicable recommendations for PR application because so many factors affect PR dissolution and resultant efficacy as a P source for crops. In general, PR use should be limited to soils with pH less than 5.5; however, PR has shown limited success on soils with pH as high as 8.0 provided there was adequate irrigation and leaching. Crop response to PR can be equal to that of commercial phosphate fertilizer on low pH soils when the PR is from a highly available source and application rates are greater than 65 kg·ha−1 P. Crop response to PR at low application rates is often less than that observed with conventional P fertilizers. Dann et al. found wheat yield increased with single superphosphate applications up to 40 kg·ha−1 P but did not find a response to sedimentary PR applications although pH was less than 5.5. Other confounding factors could have been low rainfall, PR application method (banded), and low reactivity of PR source. Low annual applications of Gafsa PR (30 kg·ha−1 per year) resulted in increased forage yields for Scholefield et al. but crop response was only half that observed from triple superphosphate [TSP; $Ca(H_2PO_4)_2 \cdot H_2O$ (0N−20P−0K)] application. Correa et al. found corn response reached a maximum at their lowest rate of TSP (75 kg·ha−1 P), but corn yield from Gafsa PR continued to increase until their maximum P application rate of 225 kg·ha−1.

Phosphate rock applications to soils with pH greater than 5.5 may require higher rates, greater amounts of incorporation, and more time to react before planting. Phosphate rock may not reach maximum solubility until 4 to 8 weeks after application. Although lime applications can benefit crop growth by reducing Al toxicity, increased pH and Ca concentrations tend to reduce the efficacy of PR. Gatiboni et al. found crop response to PR application was equal to TSP for soils with pH 5, yet combined application of PR and lime reduced yield, whereas application of TSP with lime doubled yields.

There is some evidence that cultural practices can increase the effectiveness of PR. Zaharah and Bah found that incorporation of green manures increased the dry matter yield and P uptake from soils fertilized with either sedimentary or igneous PR. Further research suggests that the decomposition products of the green manures increased the availability of P or dissolution of PR, thereby increasing P uptake from between 5% to 9% without green manure to between 19% to 48% of applied P with green manure. Furthermore, Satter et al. found that arbuscular mycorrhiza also increased P uptake from Gafsa PR-amended soils, which may have particular applicability in organically managed soils. Although PR has limited solubility and P availability is highly dependent on soil characteristics, PR may be a preferred P source in organic production of vegetable crops because, unlike manures and non-National Organic Program-compliant composts, there are not any required waiting periods between application and harvest.

Manure and Compost

In general, manures and composts are good sources of P with high plant availability. Although manures and composts are organically based nutrient sources, the majority of P present is inorganic and readily available to plants. Inorganic P accounts for 75% to 90% of the total P present in manure and compost. Unlike N, P is conserved in the composting process and, depending on the composting process, the water-soluble P of mature compost may not be different from that of the original manure source. Even compost with very low water extractable P (less than 0.01% of total P) was found to have high P availability, in which P uptake from compost-amended soils did not differ from that receiving equivalent additions of TSP. Other studies have also found that P uptake from manure and compost was equal to or greater than P uptake from commercial P fertilizers.

Although P availability of manure- or compost-based P may be as great as or greater than that of commercial fertilizers, the soil P reactions are not the same. Increases in soil test P, or measures of plant-available P, are often less for manure-based P additions than for equivalent fertilizer P additions. Griffin et al. found that that calcium chloride ($CaCl_2$)-extractable P increase was greatest with monopotassium phosphate (KH_2PO_4) addition compared with beef, poultry, swine, and dairy manures. Furthermore, applications of dairy manure up to 800 mg·kg^{-1} P soil resulted in decreased $CaCl_2$-extractable P. Research has shown that the increase in soil-test P after manure additions is inversely related to the C:P ratio of the manure. Increases in soil-test P were also inversely related to microbial biomass P, leading to the hypothesis that increased C additions stimulated microbial activity and P immobilization. Because manure-based P has less of an impact on soil test P than does commercial fertilizer, it is generally suggested that manure- and compost-based P should be considered as 70% available for soils with low soil-test P but 100% available for soils testing adequate or high for P.

Other Phosphorus Sources

There are other nutrient sources available for use in organic production that can be good sources of P. Bone meal, prepared by grinding raw animal bones, is one of the earliest P sources used in agriculture. Although bone meal is often cited as an organically approved P source, it has a relatively high cost, there are limited supplies, and research on its efficacy is limited. The primary calcium (Ca)–phosphate mineral in bone material is calcium-deficient hydroxyapatite [Ca_{10}–x-$(HPO_4)x(PO_4)_6$–x $(OH)_2$–x (o < x < 1)], which is more soluble than PR but much less soluble than conventional P fertilizers. Calcium-deficient hydroxyapatite present in bone meal would dissolve in soils as follows:

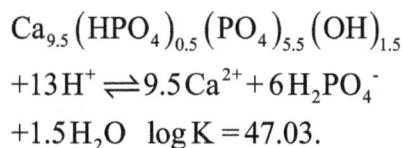

$$Ca_{9.5}\left(HPO_4\right)_{0.5}\left(PO_4\right)_{5.5}\left(OH\right)_{1.5}$$
$$+13H^+ \rightleftharpoons 9.5\,Ca^{2+}+6H_2PO_4^-$$
$$+1.5\,H_2O \quad \log K = 47.03.$$

As can be seen in Eq. 3, increasing Ca concentrations or increasing pH would decrease P release from bone meal. By fixing Ca^{2+} activity at 10–2.5 mol·L^{-1}, the phosphate concentration can be determined as a function of pH as follows:

$$\log\left(H_2PO_4^-\right)=11.8-2.17\,pH$$

From Equation (above), one can calculate that bone meal could maintain P concentration of greater than 1800 mg·L^{-1} at pH 6, but could only maintain P concentration of 0.08 mg·L^{-1} at pH 8.

Available research has shown that bone meal applications can increase crop growth and crop P uptake. Phosphorus availability can be equal to or greater than that of TSP and P availability increases with decreasing particle size. Residual effects of bone meal are equal to or greater than that of TSP. There are not many studies that report soil effects on P availability from bone meal, but it is generally recommended for use on acid soils.

Recent concerns with bovine spongiform encephalopathy (BSE) in cattle has raised some concerns about use of bone meal as a fertilizer (B. Baker, personal communication). Although it is conceivable that BSE could be transmitted from raw bone meal containing nerve tissue of infected animals to humans through soil particles on unwashed vegetables, there are no restrictions on the use of bone meal by organic farmers in the National Organic Program rules at this time., there are not any studies defining the fate of BSE prions in soils. In addition, most commercial bone meal products have been heat-treated to the point that any nerve tissue is ashed and the possibility of prion transmission has been eliminated.

Guano, more commonly known for its use as an N fertilizer, can also be used as a P source. Guano is formed from continual deposition of bird or bat droppings beneath roosting sites. As opposed to manure, guano is aged and contains various minerals that concentrate inorganic forms of nutrients. For example, struvite, a magnesium ammonium phosphate mineral, has been identified as a primary component of guano. Struvite has been used as a slow-release N and P source in the horticultural industry. Struvite can also be precipitated out of swine lagoon liquid. Although it is only slightly soluble, struvite dissolves in soils as a result of low NH_4^+ concentrations resulting from the nitrification process. As it dissolves, struvite supplies both N and P to growing plants. Studies report variable agronomic effectiveness of guano as a P source, whereas some studies have found guano equal to commercial P sources and other studies have found the agronomic effectiveness less than readily soluble P sources.

Environmental Issues Associated with Phosphorus Management

Principles of organic farming systems set forth by the International Federation of Organic Agriculture Movements places environmental protection as a primary objective of organic farming systems (International Federation of Organic Agriculture Movements. Proper P management is an important part of environmental protection in any agricultural system, especially so in systems that use manures or composts as nutrient sources, because P inputs to fresh water ecosystems are a primary cause of eutrophication and water quality degradation. Algal growth in most freshwater systems is P-limited; therefore, P inputs increase algal growth. After rapid algal growth, or the algal bloom, algal biomass decomposition reduces dissolved oxygen and can result in fish kills. Phosphorus inputs can also shift the dynamics of the algal community and stimulate growth of toxic blue–green algae. Furthermore, P inputs to salt water have been found to increase the growth of Pfiesteria piscicida, a toxic dinoflagellate linked to fish kills on the eastern coast of the United States. In general, P-induced algal blooms, low dissolved oxygen, and fish kills result in foul odors, reduced recreational value, reduced biodiversity, and increased treatment costs for fresh water ecosystems and water supplies. Therefore, organic farming systems should be designed to limit P losses through proper management of P inputs, cropping systems, and soil resources.

Understanding mechanisms controlling P loss is essential in determining the appropriate management strategies for reducing P losses. Phosphorus losses from agricultural systems generally occur through surface runoff and erosion followed by transport to streams and rivers in concentrated flow processes. Phosphorus is transported in both dissolved and particulate forms.

Particulate P losses are highly correlated to erosion rates and soil test P concentrations, in which increases in either soil test P or erosion rate will increase P loss. As previously discussed, the use of manures or composts to meet crop N demands overapplies P. Long-term P applications in excess of crop demand have been found to increase soil test P levels far beyond agronomic requirements, resulting in increased potential for P loss. Increased soil test P is also correlated to increased dissolved P concentrations in runoff. Even on soils with relatively low erosion rates, high soil test P can result in unacceptable P losses resulting from high dissolved P in runoff. Soils with high soil test P represent long-term risks of P loss because it may take many years before crop removal reduces soil test P to concentrations below environmental thresholds.

Particulate or dissolved P can also be lost directly from surface-applied manures or composts, sometimes referred to as "source P losses" because they are lost directly from the source. Source P losses can sometimes exceed soil P losses because of the high P solubility of some sources and reduced interaction and adsorption with soil. Excessive source P losses have also been observed regardless of soil test P levels. Therefore, any surface application of manure or compost should be regarded as an increased risk for P loss.

Reducing Phosphorus Losses

The first step in controlling P loss from an agricultural system is evaluating the risk of P loss under current management. Evaluation of P loss risks will help identify P sources and P transport pathways, which may be important targets for best management practices (BMPs) aimed at reducing P loss. Phosphorus loss indices, conceptualized by Lemunyon and Gilbert, have been developed and used in 47 states in the United States as well as a few European Union countries. Phosphorus loss indices the rate the relative risk of P loss by evaluating field-specific P sources and transport factors and assigning each a P risk rating. Most state-specific P indices are designed such that in the process of evaluating P loss risk, the user also identifies the major contributing factors to P losses. Once these factors have been identified, BMPs can be selected to address the specific risks.

There are a variety of best management practices that can be used to reduce P loss from agricultural systems. Because P is strongly adsorbed to soil particles, erosion control is one of the primary ways to reduce P losses. A few of the more common erosion control BMPs are contour farming, terracing, reduced tillage, grassed waterways, and cover crops, each of which will also reduce the risk of P loss. Although BMPs that stop erosion before it begins are preferred, BMPs designed to trap eroded sediments as they leave the field such as grassed buffer strips are also effective at reducing particulate P losses. Because organic farming systems do not use herbicides, continuous no tilling may not be a feasible BMP; however, short-term no tilling, rotation with cover crops, grassed waterways, and field buffers may be particularly well suited to organic farming.

Although erosion control methods may reduce the majority of P losses, they may not reduce and may even increase dissolved P losses. Dissolved P losses can be reduced by increasing infiltration rates, thereby reducing runoff volume. Infiltration can be increased through proper residue management and maintaining good soil structure. Increasing soil organic matter, one objective of organic production, has been found to increase infiltration rates.

Maintaining moderate soil test P concentrations or reducing high soil test P concentrations can reduce P lost from both erosion and dissolved P in runoff. Balancing P inputs with crop removal is an essential part of a long-term sustainable solution to controlling P losses. This may require producers who currently use animal manure or compost as the primary N source to switch to alternate N sources such as use of legumes as a green manure. Continual crop removal of P in absence of P inputs will reduce soil test P; however, reduction of soil test P through crop uptake may require many years.

Phosphorus application BMPs would be an important part of environmentally sound P management in organic agriculture because of the frequency of manure or compost additions. One of the most effective methods of reducing source P losses is to incorporate the P source into the soil through tillage immediately after application. Incorporation reduces the P source interaction with runoff water and increases P source interaction with soil, both of which tend to reduce P losses. Phosphorus application timing can also have a strong impact on source P losses. Phosphorus applications immediately followed by rainfall tend to result in greater P losses than when several days or weeks separate the application from precipitation events. Application timing BMPs are specific to climatic regimes, but in general, manure and compost applications should be made during periods with low risk of runoff being generated from either rainfall or snow melt.

References

- Types-of-organic-fertilizers: agrifarming.in, Retrieved 28 February, 2019

- Commercial-manufacturing-of-organic-fertilizer: lovetoknow.com, Retrieved 18 April, 2019

- Organic-liquid-fertilizer: bettervegetablegardening.com, Retrieved 7 May, 2019

- Liquid-organic-fertilizer: ecomena.org, Retrieved 21 July, 2019

- Composttea: homecompostingmadeeasy.com, Retrieved 15 January, 2019

- Benefits-of-the-organic-liquid-fertilizer, our-story: organicliquideco.com, Retrieved 21 April, 2019

- Poultry-litter, housing-environment, husbandry-management, production: poultryhub.org, Retrieved 15 June, 2019

- Organic-manures-meaning-classification-and-reactions, organic-manures: soilmanagementindia.com, Retrieved 18 May, 2019

- Neem-cake-gardening: gardening-abc.com, Retrieved, 17 February, 2019

- Orgfarm-vermicompost, org-farm: tnau.ac.in, Retrieved, 24 March, 2019

- Organic-phosphorus-fertilizers: grow-it-organically.com, Retrieved 1 August, 2019

Chapter 5

Organic Methods for Disease Management

Diseases in plants can be caused due to pathogens such as fungi, bacteria and viruses. There are various organic methods for disease management such as planting resistant cultivars, exclusion and crop rotation. All these diverse organic methods for disease management have been carefully analyzed in this chapter.

Plant Diseases

Looking at the spectrum of potential pathogens, or disease organisms, it's helpful to divide them into three groups:

- Fungi: Grow on or through plants via thread-like mycelium. Fungi require either living plant hosts or decaying organic matter to survive. Fungal pathogens are the greatest challenge in our region.

- Bacteria: Single-cell organisms that need a living host to survive. Bacteria reproduce readily when they have warm, moist environments and a host plant to feed on.

- Viruses: Sub-microscopic organisms that invade the host plant's cells and then multiply. Viruses spread via infected pest insects, known as vectors.

Fungal Diseases

Mildews: You've seen these a thousand times. The main types in our area are:

Downy mildew effect cucurbits like melon, cucumber, squash, and pumpkin. There is also a downy mildew that plagues basil.

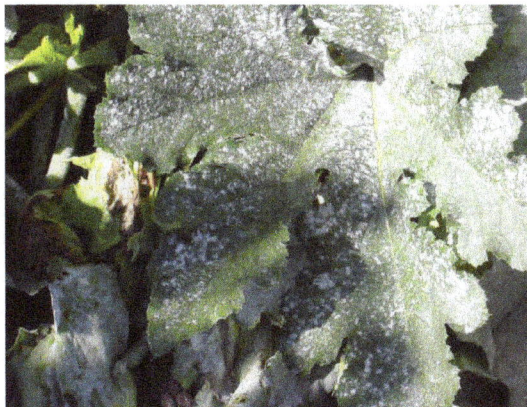

Powdery mildew is another common mildew, and can be seen on many plants both food and ornamental. In the flower garden, lilacs are susceptible, as well as roses. In the veggie garden, cucurbits often fall victim. The good news about powdery mildew is that it is species specific, meaning that the particular strains are partial to specific types of plants. So, your lilac will not give powdery mildew to your pumpkin. Both downy and powdery mildews just love stagnant, warm air and spread on the wind.

Techniques for Prevention

- Encourage good air flow.

- Plant early in the season.

- Plant disease resistant varieties.

- Provide coverage for plants using re-may.

- Plant more successions of effected annual crops like squash and cucumber, making sure to cover each succession as soon as you set it out into the garden.

Products to Consider

Biological deterrents are beneficial bacteria in powder or liquid form. They inoculate the soil and work in cooperation with plant roots to make the plant more resilient.

Check out the following:

- Bacillus subtillus (trade name serenade soil).

- Bacillus amyloliquefaciens (double nickel 55).

- Reynoutria sachalinensis (regalia).

- Calcium silicate used as a fertilizer has been shown to reduce the chances of mildew in organic cucurbit crops. OGS is still looking into this idea, and into the particular product to use, but research out of Rutgers University suggests Wollastonite powder.

Phytophthora infestans on Potato Plant

Phytophthora

Also known as "water molds" phytophthora pathogens are not actually fungi, but they closely resemble fungi, so we'll throw them in with the others. Phytophthora is what causes late blight in tomatoes, and other similar wilting diseases in almost any vegetable and ornamental crop, as well as many trees. Phytophthora is characterized by a slow wilting of the entire plant, starting at the bottom with the oldest leaves, and progressing upwards. The group of pathogens spread via spores, so wind can transmit them, as can your hands and clothes, wild animals, equipment, water, you get the drill.

Techniques for Prevention

- Prevention is key for Phytophthora, as there is little aid once the disease has established itself in your garden.

- Regulate water carefully to ensure plants are not getting too much. In seasons of overwhelming rain, this may be out of your control.

- Ensure good airflow. This is especially important in tomato crops. Keep plants pruned, space them adequately, and keep them up off of the ground.

- Keep good cropping records and rotate crops religiously.

- Choose resistant varieties, and be sure to purchase clean, healthy seed or transplants.

- Keep your hands, as well as tools and equipment sanitized while working.

- Provide a cover for plants such as a high tunnel with open sides.

- Do not prune or otherwise work your plants when they are wet.

- Don't leave debris in the garden for fungi to feed on. Remove weeds to ah ot compost pile or a burn pile away from the garden area.

Products to Consider

- Copper Sulfate is approved for organic use, and offers strong defense against fungal pathogens. Be sure to follow all safety and application instructions, as copper is a potent control method, and should be used responsibly.

- A spray regimen of serenade and copper sulfate in rotation has been effective for many small farmers in WNC. You must spray the plants thoroughly (even the undersides of leaves), make sure to spray weekly, beginning at planting and up until frost.

- Streptomyces griseoviridis (MycoStop) is a bacteria you can use to inoculate the soil. It's organic approved, and listed as a control for Phytophthora.

Septoria

Also known as leaf mold, Septoria causes brown and yellow spots on plant leaves and leads to leaf wilt. As a leaf fungus, one might think that septoria doesn't pose too much of a threat to fruiting plants, like tomatoes, however, Septoria fungi can lead to severe sun scald since the fruits are no longer shaded by leaves. Septoria is usually a problem on nightshades such as tomatoes, potatoes, eggplants, and peppers. It also affects celery. There are over 1000 species of Septoria in our region, and the fungi are known to survive on seeds.

Techniques for Prevention

- Prevention is key for Septoria, as there is little aid once the disease has established itself in your garden.

- Maintain good air flow, and for plants that are susceptible like nightshades, put some extra space between species to prevent spread of Septoria from infected plants to healthy plants. For example, don't plant peppers right next door to eggplant. Put some basil or marigolds in between.

- Remove infected leaves as you notice them and throw them in the trash (far away from the garden.)

- Don't leave debris in the garden for fungi to feed on. Remove weeds to a hot compost pile or a burn pile away from the garden area.

- Make sure your hands, as well as tools and other equipment are sanitized before working, and be careful to leave the sick plants for last so you don't carry fungal spores to healthy plants.

Products to Consider

- Copper Sulfate

- Mix 1 T horticultural oil and 1 T baking soda per gallon of water. Spray weekly as soon as you set plants out. The mix can clump so be sure to shake or stir frequently as you go.

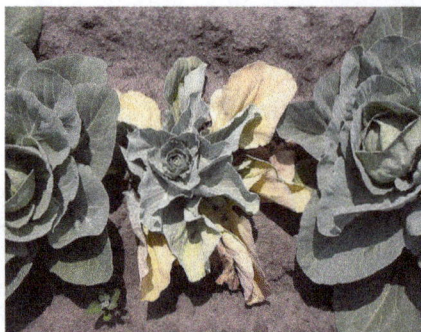

Fusarium

Wilt is a real doozie of a pathogen, causing big losses when it crops up in the garden. It can persist in the soil for years, and render planting areas virtually useless for long periods of time. It causes total wilt of plants, which can start with yellowing of lower leaves.

Techniques for Prevention

- Look for resistant varieties.

- Fusarium thrives in hot temperatures when the soil moisture is low. Be sure to keep soil evenly moist, especially in the hottest months of the season, without flooding the garden and inviting other pathogens to thrive.

- Solarizing effected soil by covering with black plastic and leaving it undisturbed during the warm season can kill fungus.

Products to Consider

- MycoStop (Streptomyces griseoviridis).

- Serenade Soil (bacillus subtillus).

Bacterial Diseases

Early Blight

Early Blight is a very common, soil-borne bacteria that effects gardeners in WNC. It causes brown spots with yellow rings on leaves and fruits. Early blight is also known as common blight, and commonly affects tomatoes and other nightshades. Unlike late blight which can kill entire plants in a day, a plant infected with early blight can persist if effected leaves are removed throughout the season. Very bad cases will result in damaged fruits.

Techniques for Prevention

- Water from below to avoid soil splashing up onto lower plant leaves. If you grow outside, this will be difficult, since rain showers cannot be controlled. If you can water from below using a soaker hose or drip irrigation AND provide a well-ventilated cover for plants to protect them from the rain, you'll be all set.

- Follow a preventative spray regimen.

- Make sure you purchase clean plant stock, from a trusted source.

- Keep equipment that you use for working the soil cleaned between uses, to prevent persistence of bacteria on your tools.

- If you do see blighty leaves (usually on the bottom of the plant closest to the soil), remove them and throw them away immediately, far from the garden.

- Do not prune, or otherwise handle your plants when they are wet.

- Establish a crop rotation and stick to it.

Products to Consider

- A spray regimen of copper sulfate and serenade in rotation has proven helpful to some farmers in WNC. Be sure to spray thoroughly (even the underside of leaves), and begin a weekly spraying from planting until frost.

Soft Rot

Soft Rotis characterized by mushy soft spots in underground crops like onions, sweet potatoes, garlic, and potatoes.

Techniques for Prevention

- Don't let potato or sweet potato seeds get too cold or wet at planting (careful not to plant too early),

- Rotate crops,

- Keep soil well drained,

- Make sure you purchase clean seed or tubers and have tools and hands sanitized at planting,

- Make sure to harvest crops when they are mature, and don't leave them in the ground too long.

Bacterial Wilt

Bacterial Wilt affects cucurbits, particularly cucumbers. It causes plants to wilt and die, and is transmitted by cucumber beetles.

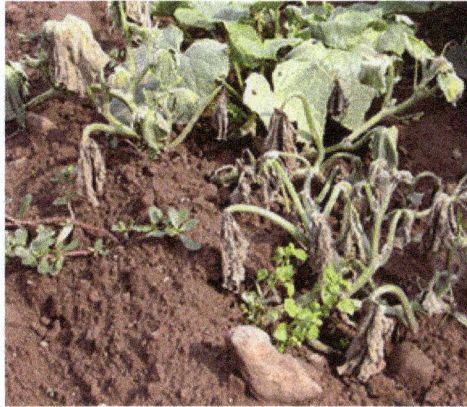

Techniques for Prevention

- Keep close control on cucumber beetle populations to prevent spread of bacteria.

Scab

Scab looks just like it sounds- rough, raised areas on the skins of underground crops, usually pota-toes. There are no varieties that are resistant to scab, and little you can do to stop it once it

happens. The best way to deal with scab is to prevent it.

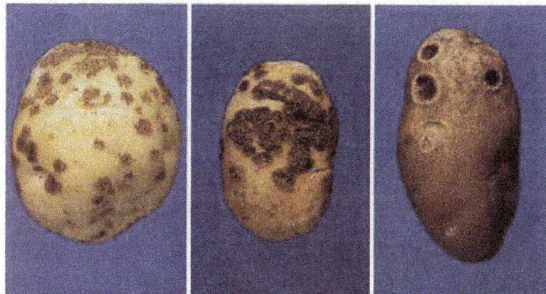

Techniques for Prevention

- Lower soil pH can be an unfavorable environment for scab. This may be unrealistic for growers with a lot of diversity and an aggressive rotation, however it may be helpful to think about planting potatoes in spring beds that have skipped a year of lime, or have had a good input of organic matter (which usually lowers pH) the previous fall

- Avoid applying lime in the spring. Look to lime your soils in the fall, as a rule.

- Purchase clean seed from a trusted source.

Viral Diseases and other Diseases

Aster Yellows is a disease that commonly effects lettuce. It is characterized by yellowing of leaves, usually beginning at the veins. It can lead to stunted growth and twisted leaves. It is spread by leafhoppers.

Techniques for Prevention and Products to Consider

- Control leafhopper populations with hot pepper spray or garlic oil soap spray. See our post on organic insect control for more info and spray recipes.

Cucumber Mosaic Virus

Cucumber Mosaic Virus, also known as Tobacco Mosaic virus, affects nightshades and cucurbits with mottled, yellow spots that look like a mosaic on the leaf surface.

Techniques for Prevention

- Avoid tobacco use around plants, and if you do smoke or chew tobacco, wash your hands thoroughly before handling plants.

- Cucumber beetles are vectors of this virus, meaning they can transmit it from plant to plant. Control cucumber beetles to prevent this disease.

- Make sure you clear weed debris from the garden area and sanitize tools and other equipment between uses.

First, let's explore some good, preventative measures that you should always try to take. Every season, re-visit this list and try to improve.

1. Strive for healthy soil with lots of organic matter, which will provide good even moisture and good drainage, as well as plenty of nutrients that plants need to stay healthy.Note: soil building will be a goal that lasts throughout your garden career. It takes years, and should be considered an investment.

2. Maintain good airflow between plants, by ensuring adequate spacing, minimal weeds, and varied architecture (i.e have tall and short plants together). Pathogens love stagnant, hot air. The better the air circulation, the better your chance of avoiding infection.

3. Water enough but not too much- Most pathogens thrive in moist to wet environments, especially as the weather heats up. Make sure you water enough to meet the requirements of your crop plants, but be especially careful about stagnant water in the garden, and plants that sit at the bottom of the garden that might collect runoff after heavy rains. Going back to #1, the healthier the soil, the better drained it will be, which will aid in your attempts at optimum water balance.

4. Look for disease resistant varieties of veggies. Note that many hybrid vegetables are bred to resist diseases that are known to affect that particular crop plant.

5. Be careful with your hands and tools because many pathogens can spread in water, on tools, on your hands, on clothes, hats, water hoses, etc. The cleaner you keep everything, the better. Also, if you suspect a group of plants have an infection of some sort but you haven't ruled out nutrient deficiency and don't want to remove them completely, remember to wash your hands after handling them before handling healthy plants. This will prevent the spread of disease in the garden.

6. Keep garden beds free of decaying debris like weeds you've pulled, or leaves you've stripped during harvesting. Fungi and bacteria like to grow on decaying organic matter. There will always be some decaying organic matter in an organic garden, but the less you contribute the better. Take weeds to your compost (only if you manage an active, hot pile) or your burn pile. Be especially aware of weeds that are in the same plant family as your crop plants, such as black nightshade (related to potatoes, tomatoes, eggplants, and peppers), as diseases that love a specific family of food plants will often get their start via weeds from the same plant family.

7. Rotate crops- Changing the planting area of crops every season will help prevent disease, especially soil-borne pathogens. Rotating your crops by family will provide extra protection. For example, if all your nightshades are in one area this season, make sure to put them as far away from that area as possible next season.

Plant Disease Management for Organic Crops

Plant diseases create challenging problems in commercial agriculture and pose real economic threats to both conventional and organic farming systems. Plant pathogens are difficult to manage for several reasons. First of all, plant pathogens are hard to identify because they are so small. The positive identification of a pathogen often requires specialized equipment and training, and in some cases accurate diagnosis in the field is difficult.

Plant pathogens are constantly changing and mutating, resulting in new strains and new challenges to growers. Also, given the local, regional, and international movement of seed, plant material, and farming equipment, new and introduced pathogens periodically enter the California system to cause new disease problems.

Disease management is complicated by the presence of multiple types of pathogens. For any one crop the grower must deal with a variety of fungi, bacteria, viruses, and nematodes. This situation is even more complicated for organic vegetable growers because they usually produce a wide array of vegetable crops and are prohibited from applying conventional synthetic fungicides. The world market continues to be extremely competitive and continues to require that growers supply high-quality, disease-free produce with an acceptable shelf life. Disease management is therefore a critical consideration in organic vegetable production.

In an organic system, it is appropriate to develop disease-control strategies that have an ecological basis. For example, insofar as possible the organic system should encourage the growth and diversity of soil in habiting and epiphytic (plant surface dwelling) microorganisms that have the potential to exert beneficial and pathogen-antagonistic influences. An increase in the genetic diversity of the crop host rotation is another management step that incorporates ecological considerations. The integration of disease management decisions with insect and weed control and general production practices is another step consistent with this approach.

Resistant Plants and Cultivars

One of the most important components in an integrated disease control program is the selection and planting of cultivars that are resistant to pathogens. The term resistance usually describes the plant host's ability to suppress or retard the activity and progress of a pathogenic agent, which results in the absence or reduction of symptoms. However, it is important to clearly establish a common definition of the term when discussing this quality with individuals from different sectors of the agricultural industry. Growers, researchers, plant breeders, and seed sellers may have slightly different understandings of the term. In addition, the word tolerance, which has a slightly different meaning, is sometimes used interchangeably with resistance, resulting in some confusion. By definition, tolerant plants can endure severe disease without suffering significant losses in quality or yield; however, these tolerant plants do not significantly inhibit the pathogen's activity, and disease symptoms may be clearly evident. Resistant plants usually suppress the pathogen in some fashion.

There are some distinct advantages to planting disease-resistant plant cultivars. Such selections are completely non-disruptive to the environment, and in fact their use enables growers to reduce and in some cases eliminate the application of chemicals used for pathogen control. The use of cultivars resistant to one disease is compatible with disease management steps taken to control other diseases. A final advantage is that for some host-pathogen systems the stability of the resistance is long lasting and the cultivars can remain resistant for many years.

There are some disadvantages to the use of resistant cultivars. The greatest shortcoming is that resistance is not available for all diseases on all crops. For several of the most damaging plant diseases, such as tomato late blight (Phytophthora infestans) and white rot (Sclerotium cepivorum) of Alliums, acceptable no resistant cultivars are yet available. In addition, commercial seed companies and plant breeders rarely invest in efforts to develop resistant cultivars for specialty or minor crops. Hence there will be specialty commodities, many of which are popular choices for organic producers that will continue to lack resistance to their disease problems. Even if resistant varieties are being developed, the long development time and high market demand for resistant cultivars will result in expensive seed, and that will affect farmer budgets.

Another shortcoming of some resistant cultivars is that some of these selections lack adequate horticultural characteristics in regard to appearance, quality, color, yield, and other important criteria. Celery resistant to Fusarium oxysporum f. sp. apii may not succumb to this Fusarium yellows fungus, but it may also be unacceptably ribby, short, or low-yielding. A cultivar that is resistant to one disease may be quite susceptible to another important disease or insect pest. A lettuce cultivar that is resistant to lettuce mosaic virus may be quite sensitive to corky root disease (caused by Rhizomonas suberifaciens); a lettuce selection that resists corky root may be very susceptible to downy mildew (Bremia lactucae). A final disadvantage to resistance is that, depending on the host-pathogen system, resistance is not long-lasting and new strains of the pathogen readily develop, making the crop susceptible once again.

Depending on the particular disease involved, the failure of plant resistance can be either a rare or a regular event. In most cases, resistance failure is attributed to the development of new strains of the target pathogen that overcome the resistance genes of the previously resistant cultivar. The downy mildew disease of spinach provides a good case study of this phenomenon. During the past 50 years in California, a new race of the spinach downy mildew fungus (Peronospora farinosa f. sp. spinaciae) would periodically occur in the state, causing significant damage to the previously resistant spinach cultivars. Plant breeders would counter with new cultivars with genes resistant to the new race. Growers would then enjoy several years of mildew-free spinach until yet another race developed. This back-and-forth dynamic has occurred for every one of the six races of the disease that have been confirmed in California.

Despite the challenges of developing resistant cultivars and the setbacks of resistance breakdown, resistant plants remain among the most important weapons for disease control in organic systems. Organic growers are encouraged to actively and thoroughly investigate which resistant cultivars are available and to test to determine which cultivars perform best under their particular growing conditions.

Site Selection

Before planting crops, a grower should carefully plan out planting and crop rotation strategies to avoid insofar as possible any known problem areas. A grower can incur significant losses if he or she plants susceptible crops in a field known to be infested with persistent soilborne pathogens. Plant-pathogenic fungi such as Armillaria, Fusarium (the wilt-causing species), Plasmodiophora, Sclerotium, and Verticillium are true soil inhabitants and will persist in soil for many years, even in the absence of a plant host. Because not all fields are infested with these fungi, the grower is advised to select a planting site away from such fields. Soilborne fungi such as Phytophthora, Pythium, and Rhizoctonia often are much more widespread, so site selection might be less of an option in avoiding these organisms.

There are also other planting situations that create risks that should be avoided. Pastures, foothills, riverbanks, grasslands, and other areas that support weeds and natural vegetation often are reservoirs of pathogens that cause virus and viruslike diseases. The vectors that carry such pathogens also can be found in these high-risk areas and often migrate into production fields. For example, the aster yellows phytoplasma and its leafhopper vector can be found in weedy grasslands in coastal California. Once the grassland vegetation dries up in the summer, the leafhoppers migrate into nearby lettuce or celery fields, resulting in aster yellows disease in these fields.

Consider pertinent environmental factors when selecting a planting site. Crops planted very close to the seacoast tend to be more at risk from downy mildew diseases as a result of the increased and persistent humidity. Just a few miles inland from the ocean, however, humidity can be significantly lower, decreasing the disease pressure for downy mildew. An understanding of soil factors is critical in avoiding some root and crown diseases. A site that has well-drained, sandy soil reduces the risk of damping-off and root rot for sensitive crops such as spinach.

For any site selection decision, careful and detailed record keeping is essential. As a grower, you should keep notes on previous soilborne disease problems associated with certain fields, the position of fields in relation to other key areas (weed reservoirs), the environmental characteristics of importance for each location, and the nature of soil, water, and other physical features of each site.

Exclusion

The practice of keeping out any materials or objects that are contaminated with pathogens or diseased plants and preventing them from entering the production system is known as exclusion. For some diseases, seed borne pathogens are a primary means of pathogen dissemination. Growers should purchase seed that has been tested and certified to be below a certain threshold infestation level or that has been treated to reduce pathogen infestation levels. Note that the designation "pathogen free seed" really is not a valid term because it is not possible to know whether a seed lot is, in its entirety, absolutely free of all pathogens. Seed tests only examine representative samples, but in most cases the tests are accurate enough to give a true picture of the risk of diseases initiated by seed borne pathogens. If a grower produces or purchases transplants, they too should be as free as possible of pathogen contamination (where the pathogen is present on the plant but has not yet caused visible symptoms) and from disease (where symptoms are actually visible).

For greenhouse crop production or the production of transplants, all materials should be clean and free of pathogens. By using clean or new pots, trays, and soilless potting mix, a grower can prevent the introduction of soil borne pathogens into the greenhouse system. The recycling of potting mixes is strongly discouraged, and pots and trays should be reused only if they are properly cleaned with steam, bleach, or other disinfectants.

Soil and water can harbor pathogens as well. Take care to see that no infested soil or water is introduced into un-infested areas. Tomato bushy stunt virus of lettuce, tomato, and other crops is found in river, flood, and runoff waters. Growers who have dredged up soil from ditches and dispersed it onto fields have found that their fields can become infested with the virus and subsequent plantings can be severely diseased. Water draining from fields can carry a number of pathogens, and growers should not recycle or reuse it without carefully considering potential risks and then taking appropriate safety precautions. Soil adhering to tractor equipment and implements can spread soilborne pathogens from infested fields into clean fields. It is a good idea to reduce the off-site movement of these infested materials as much as possible.

Applying Control Materials

Once vegetable crops are in the field or greenhouse, it will sometimes be necessary to apply some sort of protectant or eradicant spray or dust material for disease control, if one is available. Unfortunately, the selection of effective, proven materials approved for organic use is limited.

Inorganic disease control materials, primarily copper- and sulfur-based fungicides, have been used for centuries. These inorganics are generally inexpensive and widely available, and they constitute minimal threats to the environment. However, their efficacy for disease control varies. While protectant copper fungicides have some activity against a wide range of fungal and bacterial pathogens, they are not extremely effective, and sole reliance upon them probably will not result in excellent disease control. Sulfurs also exhibit some activity against many pathogens, but they usually provide excellent control against only certain pathogens, such as powdery mildew fungi. Both coppers and sulfurs can burn sensitive vegetable crops under some environmental conditions.

Oils, plant extracts, and other natural plant products are being investigated for use as disease-control sprays. Such products should be compatible with organic production practices, but reliable disease control has yet to be demonstrated.

Bicarbonate-based fungicides have recently become available for control of plant diseases. Bicarbonates have demonstrated acceptable activity against powdery mildew and a few other diseases. It is not known, however, whether bicarbonates alone will provide seasonlong protection for an organically grown crop.

Disease control using microorganisms (biocontrol) or chemical by-products made by microorganisms is generating a good deal of interest. However, the history of successful biological control of plant diseases is not encouraging. Very few effective, economically feasible biological control materials are commercially available. Much research and development remains to be done.

For the best results possible with any of these materials, appropriate application technique (proper equipment, spray volume, and plant coverage) and timing are essential. Most materials do not perform well if the disease is established, so applications should be made prior to extensive infection. Before applying a product, a grower should confirm that the material is approved for use in organic production. Consult product labels, UC Cooperative Extension farm advisors, pest control advisers, and your local Agricultural Commissioner's Office for product use information and restrictions.

Cultural Practices

There are a number of cultural practices that a grower should consider when designing an integrated disease control system. As a general approach, growers should take steps to grow vigorous, high-quality plants using the best farming practices possible. Listed below are some specific cultural practices that can help to manage diseases.

Crop rotation is an important consideration in disease management. Rotation using diverse crops, inclusion of cover crops, and appropriate use of fallow (hostfree) periods all can contribute to the reduction of inoculum levels for soilborne pathogens and the increase of diversity in soil microflora. In contrast, consecutive plantings of the same crop in the same field often lead to increases in soilborne pathogens. Too little crop rotation in a given area can also simulate a monoculture effect that might increase foliar diseases.

Recent research has shown that certain plants, besides being revenue-generating crops, also have a suppressive effect on diseases. For example, after broccoli and other crucifer crops are harvested and the plant residue is plowed into the soil, the decomposition of the broccoli stems and leaves releases natural chemicals that can significantly reduce the number of Verticillium dahliae microsclerotia. This broccoli effect can be an important consideration in crop rotation strategies.

Some cover crops (mustards, sudangrass) might also share this beneficial effect and could be consider in the crop rotation scheme. It is important to remember that while rotations with non-susceptible plants and cover crops may help reduce soilborne pathogen numbers, significant decreases in such populations are likely to take many seasons.

When devising a crop rotation strategy, a grower should also be aware of which crops and cover crops might increase disease problems. A vetch cover crop, if planted into a field with a history of lettuce drop, can greatly increase the number of infective sclerotia of Sclerotinia minor. Vetch is a known host of root-knot nematode (Meloidogyne species) and also might increase soil populations of Pythium and Rhizoctonia dampingoff fungi. While oilseed radish could be a potential trap crop for cyst nematode (Heterodera species), as a cover crop it is a host of root-knot nematode and the clubroot fungus (Plasmodiophora brassicae).

There are many factors to consider in regard to planting a crop. Timing can be an important question. If cauliflower is planted into Verticillium-infested fields in the spring or summer, it is likely to experience disease and possible crop loss. However, if cauliflower is planted into the same fields in the late fall or winter it will exhibit no Verticillium wilt symptoms, presumably because the soil temperatures are too cool to allow the fungus to develop and cause significant disease.

Disease can also be influenced by steps taken prior to and during the planting process. Tillage procedures should reduce plant residues left from previous crops. Proper preparation of the field and the subsequent raised beds should reduce problems in areas that are subject to poor drainage, pooling of water, and other conditions that favor pathogens. Soil and bed preparation should result in good soil tilth so that seed or transplants are placed in a soil that favors plant development. Planting depth for seed or transplants should be tailored to enhance seed emergence or transplant establishment. Poor soil preparation can result in stressed and exposed plants and increased damping-off problems due to soil fungi.

Irrigation management is clearly an important factor when it comes to disease control. Regardless of the irrigation method a grower chooses (furrow, sprinkler, or drip), timing and duration of irrigations should satisfy crop water requirements without allowing for excess water. Overwatering greatly favors most soilborne pathogenic fungi. For most foliar diseases, overhead sprinkler irrigation enhances pathogen survival and dispersal and disease development. Bacterial foliar diseases are particularly dependent upon rain and sprinkler irrigation. A grower should consider limiting or eliminating sprinkler irrigation if foliar diseases are problematic for a specific crop or field.

The selection and application of fertilizers, in a few documented situations, can significantly influence disease development. For example, the use of the nitrate form of nitrogenous fertilizers can increase the severity of lettuce corky root disease. The excessive use of nitrogen fertilizers can result in leaf growth that is overly succulent and more susceptible to some diseases. On the other hand, liming the soil to raise pH levels can reduce symptom expression for clubroot disease of crucifers. In general, however, fertilizer management is not directly related to disease control.

Field sanitation is the removal or destruction of diseased plant residues. In some field situations, sanitation is an appropriate step for managing diseases. Once lettuce has been harvested, for example, the remaining plants can act as a reservoir for lettuce mosaic virus. Sanitation in this case would include plowing down the old plants. Lettuce drop, caused by the fungus Sclerotinia minor, occurs when sclerotia develop on lettuce plant residues and remain in the top few inches of soil. One form of sanitation involves deep plowing in which moldboard plows invert the soil and bury sclerotia. Note that this procedure is effective only if sclerotia are low to moderate in number.

Sanitation measures are more commonly applied in greenhouse situations. The removal of dead or dying transplants can help reduce inoculum that could otherwise spread to adjacent transplants. The removal of senescent tomato or cucumber plants might reduce (though not prevent) the spread of Botrytis spores. Roguing is a special form of plant sanitation that involves the physical removal of diseased plants from the field. While not applicable in many situations, researchers have shown that for sclerotia-forming fungi (such as Sclerotinia minor on lettuce) the regular removal of diseased plants can gradually reduce the overall number of sclerotia in fields.

The management of other pests is a cultural control that could greatly influence the development of plant diseases. In particular, virus disease management is more effective when weeds and insects are also controlled. Weeds are known reservoirs of a number of viral and bacterial pathogens.

Soil solarization is the use of plastic tarps placed on the soil surface to increase soil temperatures to a level that kills soilborne pathogens, weeds, and other crop pests. Soil solarization works best in areas with acceptably high summer temperatures. These temperatures generally do not occur in California's coastal regions. Soil solarization will not eradicate a pathogen from a field, but it may lower pathogen populations. Soil flooding is a related though seldom-used means of creating conditions—in this case, saturated soil over an extended period—that might result in a decline of soilborne pathogens.

Finally, the ability to manipulate environmental conditions in a greenhouse vegetable transplant or production system can be used to help control diseases. Botrytis diseases can be better managed if warm, humid air is vented out of the greenhouse. Because rain is not a factor in greenhouses, many bacterial foliar diseases can be virtually eliminated if drip irrigation or sub-irrigation systems are used.

Composts

Incorporation of composts into soils is a fundamental cultural practice in organic production. Composts benefit the soil's fertility and condition in a number of ways, and also undoubtedly benefit disease management in some way. However, research studies and empirical data that clearly document any disease control benefits resulting from field-application of compost are lacking. Despite this lack of information on disease control, composts should be added to farmed soils in order to increase soil microflora diversity and populations.

Plant Disease Diagnostics

The first step in any management decision regarding disease control is to determine which diseases and pathogens are causing the problem. Accurate and timely diagnosis of plant diseases is an essential component of integrated disease control in organic and conventional systems. Disease diagnosis is enhanced when all professionals, including the grower, field personnel, pest control advisor, consultant, and extension personnel, work together to ascertain the cause of the problem. Often, field identification is impossible and samples must be submitted to a qualified laboratory for analysis.

Once a diagnosis has been determined, growers and other decision makers can settle on appropriate steps to take to manage the problem. Again, detailed record keeping will help the grower deal with the current problem and at the same time provide a database from which the grower can plan disease management steps for future crops.

Disease Management in Organic Lettuce Production

Lettuce (Lactuca sativa L.) is the world's most popular leafy salad vegetable. Various types of lettuce are cultivated across the world, primarily for human consumption of their fresh leaves. Among these types of lettuce are leaf (loose-leaf lettuce), Cos (romaine), crisp head (iceberg), butter head, stem (asparagus lettuce), and numerous others, which have limited production. Lettuce responds well to a moist, rich soil, full exposure to sun and cool weather conditions. It is typically a crop of temperate climates. It is grown everywhere where the average temperature remains between 45°F and 65°F during the growing season. Although primarily adapted to colder climates, there are summer and winter varieties that can be used in other seasons. Sandy peat and muck, deep black sandy loams and loams are the most suitable types of soil. Lettuce is slightly tolerant to acidic soils, but the ideal pH ranges between 6.0 and 6.8.

Damping-off (Rhizoctonia solani)   Downy mildew (Bremia lactucae)

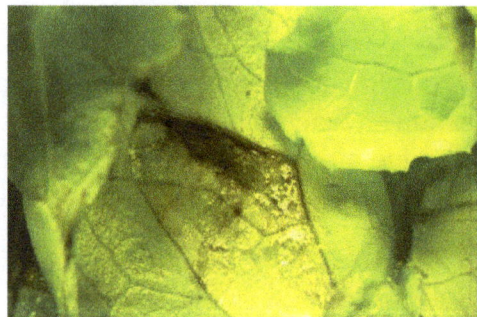

One of the major challenges facing organic lettuce producers is disease management. The losses in lettuce production due to disease can be significant and devastating under favorable conditions. The primary lettuce diseases are bottom rot (Rhizoctonia solani), damping-off (Pythium spp. and R. solani), downy mildew (Bremia lactucae), drop (Sclerotinia sclerotiorum and S. minor), gray mold (Botrytis cinerea), powdery mildew (Erysiphe cichoracearum), Septoria leaf spot (Septoria lactucae), bacterial leaf spot (Xanthomonas campestris pv. vitians), soft rot (Pectobacterium carotovorum subsp. carotovorum), aster yellows phytoplasma, lettuce mosaic virus (LMV), Turnip mosaic virus (TuMV), Beet western yellows virus (BWYV) and Tomato spotted wilt virus (TSWV). Preventive measures against such diseases have priority in organic lettuce production.

Drop (Sclerotinia minor)

Downy mildew (Bremia lactucae)

Gray mold (Botrytis cinerea)

Powdery mildew

Bacterial leaf spot

Soft rot (Pectobacterium carotovorum subsp. carotovorum)

Aster yellows phytoplasma

Disease Management Strategies

Seed selection and seed health treatments

The planting of pathogen-free organically produced lettuce seed is an important first step in managing diseases. There are some seed treatments, such as hot-water sanitation or National Organic Program (NOP)-compliant protectants that can be used by organic farmers to eradicate some pathogens from seed. Hot-water seed treatments (118 °F for 30 min) can greatly decrease seedborne inoculum of some disease such as Septoria leaf spot and bacterial leaf spot. Treatments of lettuce seed with solutions of aqueous 3 to 5% hydrogen peroxide may also effectively reduce or eradicate the bacterial leaf spot pathogen from heavily infested lettuce seed. However, hydrogen peroxide treatments at a concentration of 5% reduced seed germination up to 28% compared with controls. On the other hand, small but significant reductions in germination were observed on seed lots treated with 3% hydrogen peroxide. Treatment of lettuce seed with solutions of 0.52% sodium hypochlorite for 5 min soaking time also reduced bacterial leaf spot. However, chlorine seed treatment of organic seed must be followed by a rinse with potable water. Check with your suppliers of organic seed regarding any seed treatments they apply or use to suppress seed-borne diseases. Also, growers should check with their certifying agency regarding any seed treatments that they intend to apply prior to planting.

Variety Selection

Many cultivars (varieties) of lettuce have been developed that are resistant and tolerant to specific diseases. The terms resistance or tolerance do not mean that the plant is completely immune to disease. They refer to a plant's ability to overcome, to some degree, the effect of the pathogen. Also, no variety is resistant or tolerant to all diseases. Resistant varieties should always be used in combination with other management practices.

Organic Amendments

The addition of organic matter such as good quality compost can aid in reducing diseases caused by soilborne pathogens. Organic matter improves soil structure and its ability to hold water and nutrients; it also supports microorganisms that contribute to biological control. The incidence of lettuce drop and survival rate of sclerotia of S. sclerotiorum can be reduced by the addition of stable manure (a mixture of straw, horse dung, and urine), fowl manure (a mixture of wood shavings and chicken droppings), and Lucerne hay to the field. Note: In order to be used for organic production, animal manure must be fully composted or else incorporated into the soil at least 120 days prior to the harvest of organic lettuce. Incorporating approved amendments, such as blood meal, fish meal, or feather meal, into soil can also increase the marketable yield and quality of lettuce

Propagation and Nursery Management

Lettuce can be either transplanted or direct seeded, depending mostly on cost and availability of transplants and the time of year. Use of transplants will help manage some diseases. Direct seeding depths are about 1/4 to 3/8 inch. In case of heavier soils, shallower depth is recommended to reduce the risk of damping-off disease problems. Raised beds are ideal for lettuce production. They help prevent damage from soil compaction and flooding. They also improve airflow around the plants resulting in reduced disease incidence. Over-watering or planting in poorly drained soils must be avoided to prevent root diseases and seed decay. Plants should be irrigated without applying water to the foliage, which helps reduce most foliar lettuce diseases, particularly bacterial leaf spot. In research on irrigation methods, lettuce drop incidence was significantly lower and yields were higher in plots under subsurface drip irrigation compared with furrow irrigation. In another study, subsurface drip irrigation reduced human health risks when microbial-contaminated water was used for irrigation.

Sanitation

Many pathogens survive between crops in or on the residue from diseased plants, so it is important to remove as much of the old plant debris as possible. When lettuce drop is observed in the production area, the diseased plants and soil immediately surrounding the infected plant should be removed. Weeds should be eliminated as they may harbor pathogens or serve as hosts for insects that transmit viruses, aster yellows phytoplasma and other pathogens. Frequent disinfection of harvest tools will also help prevent the spread of soft rot, post-harvest pathogens such as gray mold, and microbial contaminants that may be human pathogens.

Rotation

Crop rotation has been used successfully for disease management in different pathosystems. Hence, it has drawn increased attention as an important disease management tool in organic agriculture. A good rotation scheme is to treat members of the same plant family as a group and rotate based on groups rather than individual crops. For example, lettuce should be rotated with vine crops, tomatoes, or cole crops but not other types of lettuce, chicory or endive. It is important to note that one difficulty in using crop rotation to control some diseases is the limited availability of crops for rotation. For example, Sclerotinia on lettuce cannot be managed with rotation because most crops are hosts of Sclerotinia and the pathogen can survive for long periods between host crops. Broccoli (Brassica oleracea), used as a green manure, recently has been shown to reduce the number of S. minor sclerotia in soil.

Row Covers

Floating row covers are an excellent barrier to some early pests that vector disease, such as leafhoppers, thrips, whiteflies and aphids. Row covers are made of lightweight fabric that can be laid directly over plants or supported with hoops. The fabric needs to be secured to keep pests out. This can be accomplished by using metal staples made for this purpose. Row covers should be removed when temperatures regularly reach the high 80's.

Materials

Early detection is important to manage diseases. It is therefore important to scout plants regularly and know which diseases are present in the crop. When preventive and cultural methods for disease control are insufficient to manage a disease, National Organic Program (NOP) compliant inputs can be applied.

Before applying ANY pest control product, be sure to 1) read the label to be sure that the product is labeled for the crop and the disease you intend to control, 2) read and understand the safety precautions and application restrictions, and 3) make sure that the brand name product is listed in your Organic System Plan and approved by your certifier.

Copper oxide can be used against lettuce downy mildew disease. However, over-application can lead to copper accumulation in the soil, contamination of run-off water, and toxicity to non-target organisms. Sulfur products can provide control for powdery mildew. Biorational products represent an important option for the management of plant diseases. On lettuce, Contans has been effective in reducing the incidence of lettuce drop by as much as 50% in greenhouse crops. In another study, two applications of Contans, one at planting and one at post-thinning, controlled lettuce drop caused by S. sclerotiorum in a desert lettuce production system.

Disease Management in Organic Seed Production

Diseases can have a significant effect on production of specialty seed crops. Seed growers must pay attention to diseases that affect the vegetative growth stage of the crop, as well as those that affect the reproductive growth stages (flowering and seed formation). Some diseases, such as Verticillium wilt of spinach, become symptomatic only when the crop enters the reproductive stage; these diseases are more important to seed growers than to vegetable growers (unless the vegetable crop also has a flowering stage, e.g., tomato or potato). While vegetable growers are concerned primarily with the pathogens that affect marketable yield and quality, seed growers must also learn how to diagnose and manage seedborne pathogens and the microorganisms that affect seed quality. Pathogens usually remain viable for longer in seed than in vegetative parts of the plant or in the soil. Seeds are a major means of survival of some plant pathogens and of introducing new pathogens to a field or region.

Disease management tactics are either preventive (actions taken to avoid or reduce the likelihood of disease problems) or curative (treatments that eliminate or reduce the effects of a particular disease after it has become established). Because there are few effective curative practices available to organic farmers, organic farmers focus their disease management efforts primarily on preventive cultural practices. Such practices include planting pathogen-free seed, planting in fields of low inoculum potential and in locations with good air movement, adopting wide row spacing, orienting the crop rows to maximize air movement between rows, and tying or staking seed crops to improve air circulation and reduce humidity in the canopy. If feasible, consider using drip or furrow irrigation instead of overhead irrigation, or irrigate earlier in the day to allow the canopy to dry before nightfall.

Some significant pathogens of seed crops are soilborne, such as Fusarium wilt of spinach. To manage soilborne pathogens, it is important to know the cropping history of the field and to adopt appropriate crop rotations. A rotation of 6 to 15 years, depending on the susceptibility of the spinach cultivar, is required to control Fusarium wilt in spinach seed crops. Some soilborne pathogens affect more than one crop, e.g., the fungus that causes Verticillium wilt of spinach can also infect potato, so it is important to avoid growing other crops in the rotation that may be alternative hosts to soilborne pathogens that affect the seed crop.

Strict management of, and screening for, seedborne pathogens of vegetable crops is critical to maintaining high seed quality. Even low levels of seed contamination can cause epidemics of some diseases when infected seed is planted in the field. For example, the tolerance level for contamination of crucifer seed with the causal agent of black rot, Xanthomonas campestris pv. campestris, is 0 contaminated seeds in 10,000 to 50,000 seeds (depending on the market or country in which the seed will be distributed).

Seeds contaminated with a pathogen can be treated physically (e.g. hot water) or chemically (e.g. bleach) to destroy inoculum or reduce the incidence of infection. Some physical and chemical treatments may reduce seed quality (germination, vigor, and longevity), so it may be important to test a particular seed treatment on a small sample of seed and check for possible phytotoxicity to the seed before treating an entire seed lot. Hot water treatment can only be used on some crops, such as brassicas, carrots, tomatoes, peppers, and lettuce, but even on those crops very precise parameters must be followed for hot water treatment to avoid damaging the seed. There are a number of biological and natural disease management products coming onto the market that are approved for use on organic farms, but it must be noted that the efficacy of these biocontrol products may vary among sites, crops, and diseases, reflecting the the complexities and particulars of interactions amongst the host, pathogen and environment. Therefore, planting pathogen-free seed, when possible, is always preferable to trying to eradicate a pathogen from seed.

Keys to Disease Management in Organic Seed Crops

Three ingredients are necessary for a disease to develop in your seed crop:

- The pathogen that causes the disease must be present.

- The host plant must be susceptible (not resistant).

- The environment (weather, microclimate) must be conducive (must support infection of the plant, growth of the pathogen).

Consider all of these ingredients when developing a disease management plan.

Pathogen

- Know the pathogen. Learn about, scout for and diagnose diseases on your crops.

- Exclude the pathogen; don't let it arrive on your farm. Don't bring in diseased transplants or seed. Don't grow your crop in the vicinity of other diseased fields to prevent wind or rain dispersal of the pathogen. Don't bring the pathogen in from other fields on people or equipment. Know which seedborne pathogens are important to your crop and prevalent in your region, and ensure, through communication with your seed supplier or contractor, that the seed you are planting has been tested to be pathogen-free or has been preventatively treated.

- Avoid the pathogen. Don't plant your crop in a field that contains the pathogen, for example, a soil that contains inoculum of Verticillium or Sclerotinia. Practice appropriate crop rotation to reduce the amount of inoculum in the field before again planting a susceptible crop. Some soilborne pathogens affect more than one crop, e.g., the fungus that causes Verticillium wilt of spinach can also infect potato, so it is important to avoid growing other crops in the rotation that may be alternative hosts to soilborne pathogens. While most soilborne diseases can be managed with a 3-5 year rotation, this is not true for some very important seed crop diseases. For example, a rotation of 6 to 15 years, depending on the susceptibility of the spinach cultivar, is required to control Fusarium wilt in spinach seed crops.

- Eradicate the pathogen. Destroy diseased crop residues, cull piles, and alternate hosts to the pathogen. Scout fields regularly for early symptoms of disease development. If feasible, rogue the symptomatic plants, and remove them from the field.

- Treat seeds for the pathogen. Treat seed with hot water or other organic treatments to eliminate the pathogen Seeds contaminated with a pathogen can be treated physically (e.g. hot water), or chemically (e.g. bleach) to destroy inoculum or reduce the incidence of infection. Some physical and chemical treatments may reduce seed quality (germination, vigor, and longevity), so it may be important to test a particular seed treatment on a small sample of seed and check for possible damage to the seed before treating an entire seed lot. Hot water treatment can be used only on some crops, such as brassicas, carrots, tomatoes, peppers, and lettuce. Even on those crops very precise parameters must be followed to avoid damaging the seed. When possible, planting pathogen-free seed is alway preferable to trying to eradicate a pathogen from seed.

The Host Plant

- Therapy: Strict management of, and screening for, seedborne pathogens of vegetable crops is critical for maintainence of high seed quality. Low levels of contaminated seed can cause epidemics of some diseases when infected seed is planted. For example, the tolerance level for contamination of crucifer seed with the causal agent of black rot, Xanthomonas campestris pv. campestris, is one contaminated seed in 10,000 to 50,000 seeds, depending on the market or country in which the seed will be distributed. This level is as important when obtaining the seed lot to grow out as it is when selling a seed lot. Typically, it is the seed company's responsibility to screen seed lots for pathogens before sale. All diagnostic laboratory results should be communicated to the seed grower. Commercial seed health tests are not available for all seedborne pathogens, particularly specialty crops grown on a small scale, and for which there has been limited seedborne pathogen research.

- Host resistance: Select resistant cultivars. When available, this is the easiest and most reliable method to control seedborne plant diseases. Many seed catalogs and extension publications provide information on resistance and susceptibility of specific cultivars to common diseases.

Figure: Wild Garden Seed lettuce disease resistance trial. Frank Morton of Wild Garden Seed conducted a trial of many lettuce cultivars over several years to screen them for disease resistance and tolerance.

- Protection: Paint materials on a plant to reduce the success of the pathogen and slow down disease progress (for example, apply copper or sulfur). Organic farmers must describe in their organic system plan their rationale for applying a pest control material. Communicate with your certifier to ensure any materials applied to crops for disease control are permitted for use on organic farms. Always ensure that the product is labeled for the crop and disease of concern. There are a number of biological and natural disease management products on the market that are approved for organic production, but their efficacy is not well documented and it may vary by site. This is a reflection of the complexity of soil - climate systems.

The Environment

- Avoid environmental conditions that promote infection and disease development.

  - Use drip irrigation or only irrigate early in the morning. Plants wetted in the morning will dry quckly in the afternoon sun, reducing the length of time that leaves remain wet. During the summer, most pathogens require 8 to 12 hours of wetness to initiate an infection.

  - Weed your field and stake your plants to improve air flow through the canopy. Good airflow speeds drying of the canopy and reduces the length of wet leaf conditions.

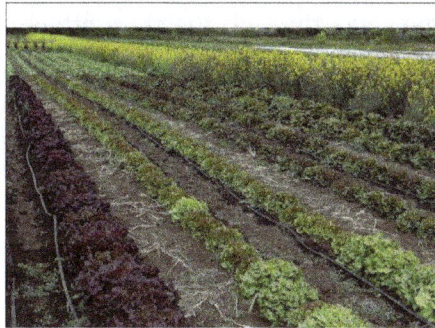

Figure: Wild Garden Seed irrigates their lettuce seed fields with drip lines and keeps them weed-free to minimize leaf wetness and disease development.

- Plant your crop earlier or later in the season to avoid weather conditions that encourage disease development. Grow your crop in high tunnels to control environmental variables.

- Use geotextile fabric to keep plant materials dry as seeds mature.

Figure: Wild Garden Seed uses geotextile fabric to keep plants and seeds dry as they mature in the field. They also use geotextile fabric to safely and easily move the seed during processing.

Early Blight Management for Organic Tomato Production

Early blight is caused by *Alternaria solani* and *A. tomatophila*, which survive between crops on infected crop residues and on solanaceous host weeds. These fungi can also be carried on tomato seed. Early blight is common on tomatoes and potatoes, and it occasionally infects eggplants and peppers. It causes direct losses by the infection of fruits and indirect losses through leaf lesions, which reduce plant vigor. The disease is favored by warm temperatures and extended periods of leaf wetness from frequent rain, overhead irrigation, or dew. The fungal spores can be spread by wind and rain, irrigation, insects, workers, and on tools and equipment. Once the primary infections have occurred, they become the most important source of new spore production and are responsible for rapid disease spread. Early blight can develop quickly mid- to late season and is more severe when plants are stressed by poor nutrition, drought, other diseases, or pests.

Symptoms

Early blight first appears as small brown-to-black lesions on older foliage. The tissue surrounding the primary lesions may become bright yellow, and when lesions are numerous, entire leaves may become chlorotic. As the lesions enlarge, they often develop concentric rings giving them a bull's-eye or target-spot appearance (Figure). When conditions are favorable for disease development, lesions can become numerous and plants defoliate, reducing both fruit quantity and quality. Fruit can become infected either in the green or ripe stage through the stem attachment. Lesions can become quite large, involve the whole fruit, and have characteristic concentric rings. Infected fruit often drops, and losses of immature fruit may occur. Fruit on defoliated plants are also subject to sunscald. Stems and petioles affected by early blight have elliptical concentric lesions, which severely weaken the plant (Fig.). Lesions at the base of emerging seedlings can cause collar rot. If this arises consecutively on many seedlings, it may indicate contamination of tomato seeds or soil used for planting.

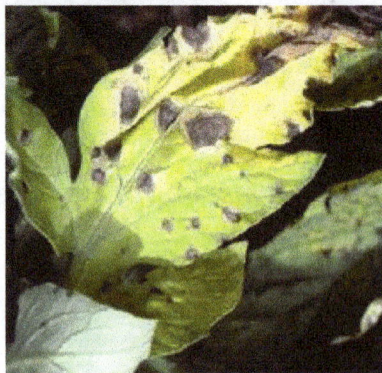

Early blight on tomato leaf

Early blight on tomato stems

Management

Seed Selection and Treatments

The planting of pathogen-free organically produced tomato seed is an important first step in managing early blight disease. Fungicidal seed treatments are not an option for organic growers; however, there are some seed treatments—such as hot water sanitation or National Organic Program (NOP)-compliant protectants—that can be used by organic farmers to eradicate some pathogens from seed. Hot water seed treatment at 122 °F for 25 minutes is recommended to control early blight on tomato seed. Chlorine seed treatment is not an acceptable treatment. Growers should always check with their certifying agency prior to using any seed treatment.

Variety Selection

A number of tomato varieties have been developed that are partially resistant to early blight, such as the Mountain series—including Mountain Pride, Mountain Supreme, Mountain Gold, Mountain Fresh, and Mountain Belle. Such varieties are sometimes inaccurately described as "tolerant" to early blight. Partially resistant varieties are not immune to a particular disease; in the case of partial resistance to early blight, infected plants may develop lesions or defoliate to some degree, often depending on environmental conditions, plant stress, and other factors.

Propagation and Nursery Management

The planting of pathogen-free organically produced tomato transplants is essential in managing early blight disease. Seeds should be sown in a pathogen-free planting mix that meets organic standards. Field soil may contain pathogens, weed seeds, and insect pests, so it is not recommended for seedling production. Flats with larger cells allow greater air movement between seedlings, which promote rapid drying of foliage and discourages disease development. Mulch (black plastic, straw, and newspaper, for example) helps to protect the plant from inoculum splashing from the soil onto lower leaves.

Crop Rotation

Early blight is a soilborne disease, so rotation can be a good management tool. A good practice is to treat members of the same plant family as a group and rotate based on groups rather than individual crops. Solanaceous crops include tomatoes, potatoes, peppers, chilies, eggplants, and tobacco. Using a three or four year crop rotation with non-solanaceous crops will allow infested plant debris to decompose in the soil. Rotations with small grains, corn, or legumes are preferable.

Organic Amendments

Good quality compost improves soil structure and its ability to hold water and nutrients; it also supports microorganisms that contribute to biological control. Our research has shown that early blight severity was less in tomato plants grown in compost-amended soil in the high tunnel than in non-amended soil; furthermore, incorporating the amendments into soil increased the total and marketable yield.

Sanitation

Early blight survives between crops in or on the residue from diseased plants, so it is important to remove diseased plants or destroy them immediately after harvest. Alternatively, bury diseased crop debris by deep-plowing to reduce spore levels available for infection of new plants. Solanaceous weeds, such as jimsonweed, horse nettle, ground-cherry, and the numerous nightshades, should be eliminated as they may harbor pathogen inoculum. Volunteer potatoes and tomatoes can also be a source of inoculum for early blight. Frequent disinfection of pruning tools should be disinfected frequently during use to help prevent the spread of spores. Stakes and cages can be disinfected each season before use with an approved product, such as ethanol, hydrogen peroxide, or peracetic acid. Disinfection with sodium hypochlorite (bleach) at 0.5% is effective, but must be followed by rinsing.

Materials

Early detection is important in disease management, so it is important to scout plants regularly. Disease forecasting is another important practice used to predict the probability of disease incidence. Weather monitoring instruments are placed in the field to collect data on canopy temperature, leaf wetness periods, and other factors that affect the likelihood of disease occurrence. The data collected from these monitoring stations are used to time fungicidal sprays for their optimum effect, generally resulting in fewer spray applications each growing season. If the uses of preventive and cultural methods for disease control are insufficient to manage early blight, National Organic Program (NOP)-compliant inputs can be applied to reduce disease spread.

Before applying any pest control product, be sure to (1) read the label to be sure that the product is labeled for the crop you intend to apply it to and the disease you intend to control, (2) read and understand the safety precautions and application restrictions, and (3) make sure that the brand name product is listed in your Organic System Plan and is approved by your certifier.

Copper products are considered synthetic and allowed with restrictions. Fixed copper products, hydrogen dioxide (=hydrogen peroxide), and potassium bicarbonate can be used against early blight. However, over-application of copper products can lead to copper accumulation in the soil, contamination of run-off water, and subsequent toxicity to non-target organisms. Biorational products represent an important option for the management of plant diseases. Research has shown that A. solani-inoculated tomato plants treated with compost extract, prepared in a ratio of 1:5 compost:water (v/v), showed a significant reduction in disease index as compared with the untreated inoculated plants. Other research has shown that the efficacy of compost tea was improved when combined with the biofungicides Serenade Max (Bacillus subtilis) and Sonata (Bacillus pumilis). These results indicate that the use of compost tea for control of tomato early blight disease may be of some benefit to greenhouse tomato growers, and to growers of organic field tomatoes who are limited in their disease management options. Garlic- and neem oils and seaweed extract have also been shown to be effective in reducing the severity of early blight disease on tomato compared to untreated controls in another research project.

Management of Black Rot of Cabbage and other Crucifer Crops in Organic Farming Systems

Black rot, caused by the bacterium *Xanthomonas campestris* pv. *campestris* (Xcc), is a significant disease of cabbage and other crucifer crops worldwide. The disease was first described in New York on turnips in 1893, and has been a common problem for growers for over 100 years. The pathogen thrives in warm, wet weather, spreading from plant to plant by splashing water, wind-blown water droplets, and by workers or animals moving from infected fields to healthy fields. Xcc can spread rapidly during transplant production in greenhouses or seed beds, and could be spreading long before any symptoms are observed. The bacterium can infest seed, infecting young seedlings as they emerge. The pathogen can also survive in cruciferous weeds, such as yellow rocket, Shepherd's purse, and wild mustard, as well as in crop debris in the field.

The cabbage above shows typical black rot symptoms, with V-shaped lesions moving into the leaf from the leaf margin

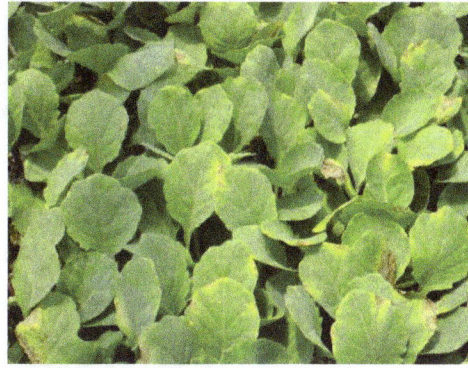

Transplants with black rot symptoms are shown above. While these plants are clearly diseased, it is important to remember that bacteria can be invading plants even if no symptoms are observed

Susceptible Crops

All crucifer crops are susceptible to black rot, including cabbage, broccoli, cauliflower, Brussels sprouts, Chinese cabbage, kale, radish, turnip, mustard, rutabaga, watercress, and arugula.

The cabbage field has been destroyed by the black rot pathogen

Symptoms and Biology

Symptoms of black rot generally begin with yellowing at the leaf margin, which expands into the characteristic "V"-shaped lesion. The bacterium commonly enters the plant through the hydathode, or water pore, on the margin of the leaf; however, damage to leaves due to insect feeding, hail, or mechanical injury can also enable pathogen entry. The bacterial infection becomes systemic, meaning that the bacterium can enter the veins of the plant and spread into the cabbage head, which can lead to serious losses in storage. Blackening of the vascular tissue is typical in severe infections.

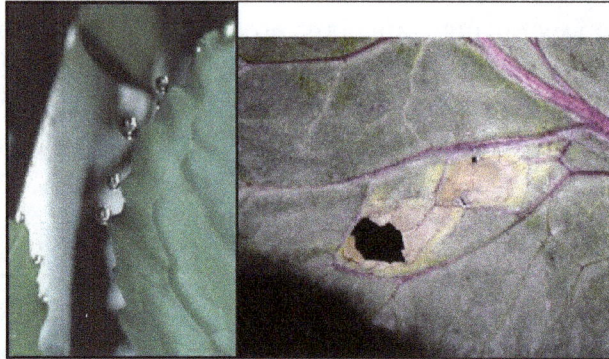

Hydathodes (or pores) on the margin of this cabbage leaf (left) exude plant sap or guttation droplets early in the morning. These hydathodes are the most common entry method for Xanthomonas campestris pv. campstris (which causes black rot). The leaf on the right is showing symptoms of black rot, with the lesion starting at the location of insect damage

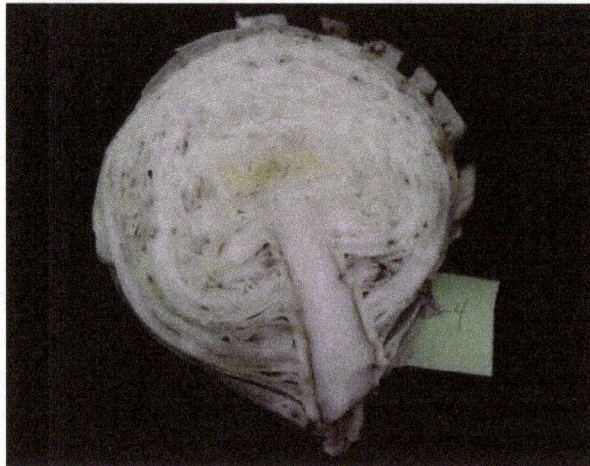

Internal vein blackening caused by the black rot pathogen. This head would rot completely during storage

Prevention

Prevention is the best line of defense and is especially important in organic production. There are three preventative measures that can reduce the risk of a black rot outbreak:

- Start with clean seed – It is known that the bacterium that causes black rot can survive on and in seed. Hot water treatment can be used to destroy the bacteria that may be infesting your seed. If you have purchased seed that has NOT been hot water treated, you can treat the seed yourself, but it is critical to do it correctly. For cabbage and Brussels sprouts, soak seed for 25 minutes in 122 °F water; for Chinese cabbage, broccoli, cauliflower, collard, kale, kohlrabi, rutabaga or turnip, soak for 20 minutes in 122 °F water. Mustards, watercress and radish are more susceptible to heat damage, and should be soaked for 15 minutes in 122°F water. Treat a small number of seeds the first time to ensure that the treatment is not reducing seed germination.

- Use clean transplants – If you are growing your own transplants, make sure that the greenhouse has been cleaned well prior to starting transplants—even if you had no disease last year! Bacteria have a remarkable way of surviving in weeds, organic matter, or nooks and crannies, so if possible, get rid of all weeds, use new or disinfected flats, and disinfect benches and tools prior to the start of a new season. Be sure to keep foliage as dry as possible, and do not brush or trim wet plants. Use pathogen-free organic starting mix, and if you are adding compost, be certain that no diseased plant matter was used.

- Rotate with non-crucifers – Because the black rot bacterium can survive in debris in the soil, it is important to rotate away from crucifer crops for a minimum of three years.

Reducing Disease Risk during the Growing Season

Anything that can be done to reduce leaf wetness and water splash will help reduce disease spread. This includes watering plants in the morning so that leaves dry prior to sunset, maintaining your irrigation system to reduce the likelihood of ponding, increasing spacing between plants, and orienting rows with prevailing winds to maximize air flow and drying.

Cabbage and cauliflower plants at this production facility are watered early
in the morning so leaves will dry quickly

Management Strategies

As with most bacterial pathogens, management can be very difficult when the weather is conducive to disease. Once a plant is infected, there is no rescue treatment since the infection is systemic. Copper-based products are effective in reducing spread from infected to healthy plants.

Before applying ANY pest control product, be sure to read and understand the safety precautions and application restrictions, and make sure that the brand name product is listed in your Organic System Plan and approved by your certifier.

Although black rot can be severe, following the prevention strategies described above will reduce the risk of this disease. Although the pathogen can survive on farms, we know that this is not the most common source of inoculum on farms that use a minimum three year rotation; instead, the pathogen is most commonly brought onto farms on seed or plants. In New York, new strains of the pathogen enter the state each year. Thus, planting only clean seed and disease-free transplants are the most important management practices in regions with cold winters. In locations with mild winter temperatures, the risk of maintaining the pathogen on farms is greater.

References

- Organic-disease-control, library, gardeners: organicgrowersschool.org, Retrieved 21 July, 2019

- Disease-management-in-organic-lettuce-production: extension.org, Retrieved 15 May, 2019

- Disease-management-in-organic-seed-production: extension.org, Retrieved 13 July, 2019

- Keys-to-disease-management-in-organic-seed-crops: extension.org Retrieved 21 April, 2019

- Early-blight-management-for-organic-tomato-production: extension.org, Retrieved 17 February, 2019

- Managing-black-rot-of-cabbage-and-other-crucifer-crops-in-organic-farming-systems: extension.org, Retrieved 24 March, 2019

Permissions

Index